Progress in Nonlinear Differential Equations and Their Applications
Volume 24

Michel Willem

Minimax Theorems

Birkhäuser
Boston • Basel • Berlin

Michel Willem
Département de Mathématique
Université Catholique de Louvain
B. 1348 Louvain-la-Neuve
Belgium

Library of Congress Cataloging-in-Publication Data

Willem, Michel.
    Minimax theorems / Michel Willem.
        p.   cm. -- (Progress in nonlinear differential equations and
    their applications ; v. 24)
    Includes bibliographical references and index.
    ISBN 0-8176-3913-6 (h : alk. paper). -- ISBN 3-7643-3913-6 (h :
alk. paper)
        1. Boundary value problems.   2.   Maxima and minima.
    3. Mathematical physics. I. Series.
    QC20.7.B6W55   1996   96-19919
    515'.64--dc20   CIP

Printed on acid-free paper
© 1996 Birkhäuser   *Birkhäuser*

ISBN 0-8176-3913-6
ISBN 3-7643-3913-6

Typeset copy prepared from author's L^ATEX disk
Printed and bound by Quinn-Woodbine, Woodbine, NJ
Printed in the United States of America

9 8 7 6 5 4 3 2 1

A Eliane

Contents

Minimax Theorems

Introduction

Many boundary value problems are equivalent to

$$(1) \qquad\qquad Au = 0$$

where $A : X \to Y$ is a mapping between two Banach spaces. When the problem is variational, there exists a differentiable functional $\varphi : X \to \mathbb{R}$ such that $A = \varphi'$, i.e.

$$\langle Au, v \rangle = \lim_{t \to 0} \frac{\varphi(u + tv) - \varphi(u)}{t}.$$

The space Y corresponds then to the topological dual X' of X and equation (1) is equivalent to $\varphi'(u) = 0$, i.e.

$$(2) \qquad\qquad \langle \varphi'(u), v \rangle = 0, \quad \forall v \in X.$$

A *critical point* of φ is a solution u of (2) and the value of φ at u is a *critical value* of φ. How to find critical values? When φ is bounded from below, the infimum

$$c := \inf_X \varphi$$

is a natural candidate. Ekeland's variational principle implies the existence of a sequence (u_n) such that

$$\varphi(u_n) \to c, \quad \varphi'(u_n) \to 0.$$

Such a sequence is called a *Palais-Smale sequence at level c*. The functional φ satisfies the $(PS)_c$ *condition* if any Palais-Smale sequence at level c has a convergent subsequence. If φ is bounded from below and satisfies the $(PS)_c$ condition at level $c := \inf_X \varphi$, then c is a critical value of φ.

Following Ambrosetti and Rabinowitz, we consider now the case when φ has at 0 a local, but not a global, minimum. There exists $r > 0$ and $e \in X$ such that $\|e\| > r$ and

$$\inf_{\|u\|=r} \varphi(u) > \varphi(0) \geq \varphi(e).$$

The point $(0, \varphi(0))$ is separated from the point $(e, \varphi(e))$ by a ring of mountains. If we consider the set Γ of paths joining 0 to e then

$$c := \inf_{\gamma \in \Gamma} \max_{t \in [0,1]} \varphi(\gamma(t))$$

is also a natural candidate. Again Ekeland's variational principle implies the existence of a sequence (u_n) such that

$$\varphi(u_n) \rightarrow c, \quad \varphi'(u_n) \rightarrow 0.$$

But c is not in general a critical value of φ.

More generally, we define

$$c := \inf_{B \in \mathcal{B}} \sup_{u \in B} \varphi(u),$$

where $\mathcal{B}$ is a class of subsets of X, and we try to prove the existence of a Palais-Smale sequence at level c. Ekeland's variational principle was used in 1984 by Aubin and Ekeland in the construction of Palais-Smale sequences. In this book, we use the more natural method of deformations. The functional φ induces a *filtration* of X defined by

$$\varphi^d := \{u \in X : \varphi(u) \leq d\}.$$

When X is a Hilbert space and $\varphi \in C^2(X, \mathbb{R})$, we deform φ^d along the *gradient flow* given by

$$\begin{cases} \frac{d}{dt}\sigma(t, u) = -\nabla\varphi(\sigma(t, u)), \\ \sigma(0, u) = u, \end{cases}$$

where

$$(\nabla\varphi(v), w) = \langle \varphi'(v), w \rangle.$$

In the general case, we use a *pseudo-gradient flow* defined by Palais in 1966. The usual minimax method consists of three steps:
1) *a priori* compactness condition, like $(PS)_c$,
2) deformation lemma depending on this condition,
3) construction of a critical value.

In 1983, the author proved a quantitative deformation lemma independent of any compactness assumption. In 1987, Mawhin used this lemma in the construction of Palais-Smale sequences. This new method consists also of three steps:
1) quantitative deformation lemma,
2) construction of a Palais-Smale sequence,
3) *a posteriori* compactness condition.

This approach is simpler and more general than the usual one. It can be applied to many problems where the $(PS)_c$ condition fails. Moreover the construction of Palais-Smale sequences is clearly separated from their compactness.

The construction of a Palais-Smale sequence depends on a *topological intersection property*. In the case of the Ambrosetti-Rabinowitz theorem, we use only the intermediate value theorem. Any path joining 0 to e intersects the sphere $S_r := \{u \in X : ||u|| = r\}$. In Chapter 2, we use the non retractibility of the ball B^N onto the sphere S^{N-1}, in Chapter 3, the Borsuk-Ulam theorem and in Chapter 6, the Kryszewski-Szulkin degree.

Chapters 3, 4 and 5 are devoted to multiplicity results. Chapter 3 contains the "fountain theorem" proved by Bartsch in 1993. This result extends the symmetric mountain pass theorem of Ambrosetti and Rabinowitz. We prove also the "dual fountain theorem". In Chapter 5, we generalize the classical category theory of Lusternik and Schnirelman. We use a variant, due to Szulkin, of the relative category defined by Reeken in 1972.

We apply some basic minimax theorems to the model problem

$$(\mathcal{P}) \qquad \begin{cases} -\Delta u + \lambda u = |u|^{p-2}u, \\ u \in H_0^1(\Omega), \end{cases}$$

or to some variants. We denote by Ω a domain of $\mathbb{R}^N$ and by $H_0^1(\Omega)$ the closure of $\mathcal{D}(\Omega)$ with respect to the norm

$$||u||_1 := \int_\Omega (|\nabla u|^2 + u^2)dx.$$

Let $\varphi : H_0^1(\Omega) \to \mathbb{R}$ be defined by

$$\varphi(u) := \int_\Omega \Big[\frac{|\nabla u|^2}{2} + \frac{\lambda u^2}{2} - \frac{|u|^p}{p} \Big] dx.$$

Since

$$\langle \varphi'(u), v \rangle = \int_\Omega [\nabla u \cdot \nabla v + \lambda uv - |u|^{p-2}uv]dx,$$

the critical points of φ are the weak solutions of $(\mathcal{P})$. In the superquadratic case,

$$2 < p < \infty, \quad N = 1, 2,$$
$$2 < p \leq 2^* := 2N/(N-2), \quad N \geq 3,$$

it is easy to verify that

$$\sup_{H_0^1} \varphi = -\inf_{H_0^1} \varphi = \infty.$$

In 1960, Nehari proved the existence of a nontrivial solution of $(\mathcal{P})$ when $\lambda \geq 0$ and $\Omega =]a, b[$, by considering

$$c := \inf_{\mathcal{N}} \varphi, \quad \mathcal{N} := \{u \in H_0^1 : \langle \varphi'(u), u \rangle = 0, u \neq 0\}.$$

In 1961, he proved the existence of infinitely many solutions and, in 1963, he solved the case where $\Omega = \mathbb{R}^3$, $\lambda > 0$ and $2 < p < 6$ after reduction to an ordinary differential equation. When Ω is unbounded or when $p = 2^*$, there is a lack of compactness because of invariance by translation or by dilation. Some nonexistence results follow from the Pohozaev identity. General existence theorems were first obtained by Strauss in 1977 when $\Omega = \mathbb{R}^N$ and by Brézis and Nirenberg in 1983 when $p = 2^*$. The Brézis-Lieb Lemma and Pierre-Louis Lions concentration-compactness method are important tools for those problems.

In Chapter 6, we consider the problem

$$(\mathcal{Q}) \qquad \begin{cases} -\Delta u + V(x)u = f(x, u), \\ u \in H^1(\mathbb{R}^N), \end{cases}$$

where V and f are periodic with respect to x_k, $k = 1, ..., N$, f is sublinear near $u = 0$ and superlinear, but subcritical, near infinity. This problem is rather difficult since the $(PS)_c$ conditions fails at every nonzero critical level. Moreover the problem is strongly indefinite since we assume that 0 lies in a spectral gap of the linear operator

$$H^2(\mathbb{R}^N) \to L^2(\mathbb{R}^N) : u \to -\Delta u + V(x)u.$$

This operator is then indefinite on any space of finite codimension. A similar problem, in the context of periodic solutions of Hamiltonian systems, was first solved by Rabinowitz in 1978. Because of lack of compactness the abstract theorem of Benci and Rabinowitz is not applicable to problem $(\mathcal{Q})$.

Some of the above problems are motivated by the existence of *solitary waves*. Consider, for example, the nonlinear Schrödinger equation

$$(\text{NLS}) \qquad\qquad iv_t + \Delta v + f(|v|)v = 0.$$

A *standing wave* is a solution of the form

$$v(t, x) = e^{i\omega t} u(x), \quad \omega \in \mathbb{R}.$$

A *traveling wave* is a solution of the form

$$v(t, x) = u(x - ct), \quad |c| < 1.$$

Also of interest are the generalized Kortweg-de Vries equation

$$\text{(GKdV)} \qquad\qquad v_t + v_{xxx} + (f(v))_x = 0$$

and the generalized Kadomtsev-Petviashvili equation

$$\text{(GKP)} \qquad\qquad v_t + v_{xxx} + (f(v))_x = D_x^{-1} v_{yy}$$

where

$$D_x^{-1} h(x, y) = \int_{-\infty}^{x} h(s, y) ds.$$

Chapter 7 is devoted to the existence of traveling waves of the generalized Kadomtsev-Petviashvili equation.

The reader interested in other aspects of critical point theory can consult the following books: [2], [6], [8], [30], [35], [38], [42], [44], [56], [57], [66], [70], [76], [79], [80], [81].

I am grateful to Allan Lazer for introducing me to minimax theory in 1980, to Thomas Bartsch, Shujie Li, Jean Mawhin, Andrzej Szulkin and Zhi-Qiang Wang for enlightening discussions. The early departure of my friend Gilles Fournier has been a sad loss. I thank the staff at Birkhäuser for their efficient editorial work. Finally special thanks are due to Suzanne D'Addato for her outstanding conversion of the manuscript into TeX.

Chapter 1

Mountain pass theorem

1.1 Differentiable functionals

Let us recall some notions of differentiability.

Definition 1.1. Let $\varphi : U \to \mathbb{R}$ where U is an open subset of a Banach space X. The functional φ has a Gateaux derivative $f \in X'$ at $u \in U$ if, for every $h \in X$,

$$\lim_{t \to 0} \frac{1}{t}[\varphi(u + th) - \varphi(u) - \langle f, th \rangle] = 0.$$

The Gateaux derivative at u is denoted by $\varphi'(u)$.

The functional φ has a Fréchet derivative $f \in X'$ at $u \in U$ if

$$\lim_{h \to 0} \frac{1}{||h||}[\varphi(u + h) - \varphi(u) - \langle f, h \rangle] = 0.$$

The functional φ belongs to $\mathcal{C}^1(U, \mathbb{R})$ if the Fréchet derivative of φ exists and is continuous on U.

If X is a Hilbert space and φ has a Gateaux derivative at $u \in U$, the gradient of φ at u is defined by

$$(\nabla\varphi(u), h) := \langle \varphi'(u), h \rangle.$$

Remarks 1.2. a) The Gateaux derivative is given by

$$\langle \varphi'(u), h \rangle := \lim_{t \to 0} \frac{1}{t}[\varphi(u + th) - \varphi(u)].$$

b) Any Fréchet derivative is a Gateaux derivative. Using the mean value theorem, it is easy to prove the following result:

Proposition 1.3. If φ has a continuous Gateaux derivative on U then $\varphi \in \mathcal{C}^1(U, \mathbb{R})$.

Definition 1.4. *Let $\varphi \in C^1(U, \mathbb{R})$. The functional φ has a second Gateaux derivative $L \in \mathcal{L}(X, X')$ at $u \in U$ if, for every $h, v \in X$,*

$$\lim_{t \to 0} \frac{1}{t} \langle \varphi'(u + th) - \varphi'(u) - Lth, v \rangle = 0.$$

The second Gateaux derivative at u is denoted by $\varphi''(u)$.

The functional φ has a second Fréchet derivative $L \in \mathcal{L}(X, X')$ at $u \in U$ if

$$\lim_{h \to 0} \frac{1}{||h||} [\varphi'(u + h) - \varphi'(u) - Lh] = 0.$$

The functional φ belongs to $C^2(U, \mathbb{R})$ if the second Fréchet derivative of φ exists and is continuous on U.

Remarks 1.5. a) The second Gateaux derivative is given by

$$\langle \varphi''(u)h, v \rangle := \lim_{t \to 0} \frac{1}{t} \langle \varphi'(u + th) - \varphi'(u), v \rangle.$$

b) Any second Fréchet derivative is a second Gateaux derivative. Using the mean value theorem, it is easy to prove the following:

Proposition 1.6. *If φ has a continuous second Gateaux derivative on V then $\varphi \in C^2(U, \mathbb{R})$.*

We will use the following function spaces.

Definition 1.7. *The space*

$$H^1(\mathbb{R}^N) := \{u \in L^2(\mathbb{R}^N) : \nabla u \in L^2(\mathbb{R}^N)\}$$

with the inner product

$$(u, v)_1 := \int_{\mathbb{R}^N} [\nabla u \cdot \nabla v + uv]$$

and the corresponding norm

$$||u||_1 := \left(\int_{\mathbb{R}^N} |\nabla u|^2 + |u|^2 \right)^{1/2}$$

is a Hilbert space. Let Ω be an open subset of $\mathbb{R}^N$. The space $H_0^1(\Omega)$ is the closure of $\mathcal{D}(\Omega)$ in $H^1(\mathbb{R}^N)$.

Let $N \geq 3$ and $2^* := 2N/(N - 2)$. The space

$$\mathcal{D}^{1,2}(\mathbb{R}^N) := \{u \in L^{2^*}(\mathbb{R}^N) : \nabla u \in L^2(\mathbb{R}^N)\}$$

with the inner product

$$\int_{\mathbb{R}^N} \nabla u \cdot \nabla v$$

and the corresponding norm

$$\left(\int_{\mathbb{R}^N} |\nabla u|^2 \right)^{1/2}$$

is a Hilbert space. The space $\mathcal{D}_0^{1,2}(\Omega)$ is the closure of $\mathcal{D}(\Omega)$ in $\mathcal{D}^{1,2}(\mathbb{R}^N)$. For simplicity of notations, we shall write $2^* = \infty$ when $N = 1$ or $N = 2$.

For the following results, see [20] or [90].

Theorem 1.8. (Sobolev imbedding theorem). *The following imbeddings are continuous:*

$$\begin{aligned}
H^1(\mathbb{R}^N) &\subset L^p(\mathbb{R}^N), & 2 \leq p < \infty, N = 1, 2, \\
H^1(\mathbb{R}^N) &\subset L^p(\mathbb{R}^N), & 2 \leq p \leq 2^*, N \geq 3, \\
D^{1,2}(\mathbb{R}^N) &\subset L^{2^*}(\mathbb{R}^N), & N \geq 3.
\end{aligned}$$

In particular, the Sobolev inequality holds:

$$S := \inf_{\substack{u \in \mathcal{D}^{1,2}(\mathbb{R}^N) \\ |u|_{2^*} = 1}} |\nabla u|_2^2 > 0.$$

Theorem 1.9. (Rellich imbedding theorem). *If $|\Omega| < \infty$, the following embeddings are compact:*

$$H_0^1(\Omega) \subset L^p(\Omega), \quad 1 \leq p < 2^*.$$

Corollary 1.10. (Poincaré inequality). *If $|\Omega| < \infty$, then*

$$\lambda_1(\Omega) := \inf_{\substack{u \in H_0^1(\Omega) \\ |u|_2 = 1}} |\nabla u|_2^2 > 0$$

is achieved.

Remarks 1.11. a) It is clear that $H_0^1(\Omega) \subset \mathcal{D}_0^{1,2}(\Omega)$.
b) If $|\Omega| < \infty$, Poincaré inequality implies that $H_0^1(\Omega) = \mathcal{D}_0^{1,2}(\Omega)$.

Proposition 1.12. *Let Ω be an open subset of $\mathbb{R}^N$ and let $2 < p < \infty$. The functionals*

$$\psi(u) := \int_\Omega |u|^p, \quad \chi(u) := \int_\Omega |u^+|^p$$

are of class $C^2(L^p(\Omega), \mathbb{R})$ and

$$\langle \psi'(u), h \rangle = p \int_\Omega |u|^{p-2} u h, \quad \langle \chi'(u), h \rangle = p \int_\Omega (u^+)^{p-1} h.$$

Proof. **Existence of the Gateaux derivative.** We only consider ψ. The proof for χ is similar. Let $u, h \in L^p$. Given $x \in \Omega$ and $0 < |t| < 1$, by the mean value theorem, there exists $\lambda \in]0, 1[$ such that

$$\big| |u(x) + th(x)|^p - |u(x)|^p \big| / |t| = p|u(x) + \lambda th(x)|^{p-1}|h(x)|$$
$$\leq p[|u(x)| + |h(x)|]^{p-1}|h(x)|.$$

The Hölder inequality implies that

$$[|u(x)| + |h(x)|]^{p-1}|h(x)| \in L^1(\Omega).$$

It follows then from the Lebesgue theorem that

$$\langle \psi'(u), h \rangle = p \int_\Omega |u|^{p-2}uh.$$

Continuity of the Gateaux derivative. Let us define $f(u) := p|u|^{p-2}u$. Assume that $u_n \to u$ in L^p. Theorem A.2 or A.4 implies that $f(u_n) \to f(u)$ in L^q when $q := p/(p-1)$. We obtain, by the Hölder inequality,

$$|\langle \psi'(u_n) - \psi'(u), h \rangle| \leq |f(u_n) - f(u)|_q |h|_p,$$

and so

$$\|\psi'(u_n) - \psi'(u)\| \leq |f(u_n) - f(u_n)|_q \to 0, n \to \infty.$$

Existence of the second Gateaux derivative. Let $u, h, v \in L^p(\Omega)$. Given $x \in \Omega$ and $0 < |t| < 1$, by the mean value theorem, there exists $\lambda \in]0, 1[$ such that

$$|[f(u(x) + th(x)) - f(u(x))]v(x)|/|t|$$
$$= p(p-1)|u(x) + \lambda th(x)|^{p-2}|h(x)|\,|v(x)|$$
$$\leq p(p-1)[|u(x)| + |h(x)|]^{p-2}|h(x)|\,|v(x)|.$$

The Hölder inequality implies that

$$[|u(x)| + |h(x)|]^{p-2}|h(x)|\,|v(x)| \in L^1(\Omega).$$

It follows then from the Lebesgue theorem that

$$\langle \psi''(u)h, v \rangle = p(p-1) \int_\Omega |u|^{p-2}hv.$$

Continuity of the second Gateaux derivative. Let us define $g(u) := p(p-1)|u|^{p-2}$. Assume that $u_n \to u$ in L^p. Theorem A.2 or A.4 implies that $g(u_n) \to g(u)$ in L^r where $r := p/(p-2)$. We obtain, by the Hölder inequality,

$$|\langle (\psi''(u_n) - \psi''(u))h, v \rangle \leq |g(u_n) - g(u)|_r |h|_p |v|_p,$$

and so

$$||\psi''(u_n) - \psi''(u)|| \leq |g(u_n) - g(u_n)|_r \to 0, \quad n \to \infty. \qquad \square$$

Corollary 1.13. a) *Let* $2 < p < \infty$ *if* $N = 1, 2$ *and* $2 < p \leq 2^*$ *if* $N \geq 3$. *The functionals* ψ *and* χ *are of class* $C^2(H_0^1(\Omega), \mathbb{R})$.
b) *Let* $N \geq 3$ *and* $p = 2^*$. *The functional* ψ *and* χ *are of class* $C^2(\mathcal{D}_0^{1,2}(\Omega), \mathbb{R})$.

Proof. The result follows directly from the Sobolev theorem. $\square$

1.2 Quantitative deformation lemma

We will prove a simple case of the quantitative deformation lemma. The general version will be given in the next chapter. Let us recall that $\varphi^d := \varphi^{-1}(] - \infty, d])$.

Lemma 1.14. *Let* X *be a Hilbert space,* $\varphi \in C^2(X, \mathbb{R})$, $c \in \mathbb{R}$, $\varepsilon > 0$. *Assume that*

$$(\forall u \in \varphi^{-1}([c - 2\varepsilon, c + 2\varepsilon])) : ||\varphi'(u)|| \geq 2\varepsilon.$$

Then there exists $\eta \in C(X, X)$ *such that*
(i) $\eta(u) = u, \forall u \notin \varphi^{-1}([(c - 2\varepsilon, c + 2\varepsilon])$,
(ii) $\eta(\varphi^{c+\varepsilon}) \subset \varphi^{c-\varepsilon}$.

Proof. Let us define

$$
\begin{aligned}
A &:= \varphi^{-1}([c - 2\varepsilon, c + 2\varepsilon]), \\
B &:= \varphi^{-1}([c - \varepsilon, c + \varepsilon]), \\
\psi(u) &:= \operatorname{dist}(u, X \backslash A)(\operatorname{dist}(u, X \backslash A) + \operatorname{dist}(u, B))^{-1},
\end{aligned}
$$

so that ψ is locally Lipschitz continuous, $\psi = 1$ on B and $\psi = 0$ on $X \backslash A$. Let us also define the locally Lipschitz continuous vector field

$$
\begin{aligned}
f(u) &:= -\psi(u)||\nabla\varphi(u)||^{-2}\nabla\varphi(u), & u \in A, \\
&:= 0, & u \in X \backslash A.
\end{aligned}
$$

It is clear that $||f(u)|| \leq (2\varepsilon)^{-1}$ on X. For each $u \in X$, the Cauchy problem

$$
\begin{aligned}
\frac{d}{dt}\sigma(t, u) &= f(\sigma(t, u)), \\
\sigma(0, u) &= u,
\end{aligned}
$$

has a unique solution $\sigma(., u)$ defined on $\mathbb{R}$. Moreover, σ is continuous on $\mathbb{R} \times X$ (see e.g. [78]). The map η defined on X by $\eta(u) := \sigma(2\varepsilon, u)$ satisfies (i). Since

$$(1.1) \qquad \frac{d}{dt}\varphi(\sigma(t,u)) = \left(\nabla\varphi(\sigma(t,u)), \frac{d}{dt}\sigma(t,u)\right)$$
$$= (\nabla\varphi(\sigma(t,u)), f(\sigma(t,u)))$$
$$= -\psi(\sigma(t,u))$$

$\varphi(\sigma(.,u))$ is nonincreasing. Let $u \in \varphi^{c+\varepsilon}$. If there is $t \in [0, 2\varepsilon]$ such that $\varphi(\sigma(t, u)) < c - \varepsilon$, then $\varphi(\sigma(2\varepsilon, u)) < c - \varepsilon$ and (ii) is satisfied. If

$$\sigma(t, u) \in \varphi^{-1}([c - \varepsilon, c + \varepsilon]), \forall t \in [0, 2\varepsilon],$$

then we obtain from (1.1),

$$\varphi(\sigma(2\varepsilon, u)) = \varphi(u) + \int_0^{2\varepsilon} \frac{d}{dt}\varphi(\sigma(t, u))dt$$
$$= \varphi(u) - \int_0^{2\varepsilon} \psi(\sigma(t, u))dt$$
$$\leq c + \varepsilon - 2\varepsilon = c - \varepsilon,$$

and (ii) is also satisfied. $\square$

1.3   Mountain pass theorem

The mountain pass theorem is the simplest and one of the most useful minimax theorems.

Theorem 1.15. *Let X be a Hilbert space, $\varphi \in C^2(X, \mathbb{R})$, $e \in X$ and $r > 0$ be such that $\|e\| > r$ and*

$$(1.2) \qquad b := \inf_{\|u\|=r} \varphi(u) > \varphi(0) \geq \varphi(e).$$

Then, for each $\varepsilon > 0$, there exists $u \in X$ such that
a) $c - 2\varepsilon \leq \varphi(u) \leq c + 2\varepsilon$,
b) $\|\varphi'(u)\| < 2\varepsilon$,
where

$$c := \inf_{\gamma \in \Gamma} \max_{t \in [0,1]} \varphi(\gamma(t))$$

and

$$\Gamma := \{\gamma \in C([0, 1], X) : \gamma(0) = 0, \gamma(1) = e\}.$$

Proof. Assumption (1.2) implies that

$$b \leq \max_{t \in [0,1]} \varphi(\gamma(t)),$$

and so

$$b \leq c \leq \max_{t \in [0,1]} \varphi(te).$$

Suppose that, for some $\varepsilon > 0$, the conclusion of the theorem is not satisfied. We may assume

(1.3) $$c - 2\varepsilon \geq \varphi(0) \geq \varphi(e).$$

By the definition of c, there exists $\gamma \in \Gamma$ such that

(1.4) $$\max_{t \in [0,1]} \varphi(\gamma(t)) \leq c + \varepsilon.$$

Consider $\beta := \eta \circ \gamma$, where η is given by the preceding lemma. We have, using (i) and (1.3),

$$\beta(0) = \eta(\gamma(0)) = \eta(0) = 0,$$

and similarly $\beta(1) = e$, so that $\beta \in \Gamma$. It follows from (ii) and (1.4) that

$$c \leq \max_{t \in [0,1]} \varphi(\beta(t)) \leq c - \varepsilon.$$

This is a contradiction. $\square$

In order to prove that c is a critical value of φ, we need the following compactness condition.

Definition 1.16. (Brézis-Coron-Nirenberg, 1980). *Let X be a Banach space, $\varphi \in C^1(X, \mathbb{R})$ and $c \in \mathbb{R}$. The function φ satisfies the $(PS)_c$ condition if any sequence $(u_n) \subset X$ such that*

(1.5) $$\varphi(u_n) \to c, \quad \varphi'(u_n) \to 0$$

has a convergent subsequence.

Theorem 1.17. (Ambrosetti-Rabinowitz, 1973). *Under the assumption of Theorem 1.15, if φ satisfies the $(PS)_c$ condition, then c is a critical value of φ.*

Proof. Theorem 1.15 implies the existence of a sequence $(u_n) \subset X$ satisfying (1.5). By $(PS)_c$, (u_n) has a subsequence converging to $u \in X$. But then $\varphi(u) = c$ and $\varphi'(u) = 0$. $\square$

Example 1.18. (Brézis-Nirenberg, 1991). *Under the assumptions of Theorem 1.15, c is not, in general, a critical value of φ. Let us define $\varphi \in C^\infty(\mathbb{R}^2, \mathbb{R})$ by*

$$\varphi(x, y) := x^2 + (1 - x)^3 y^2.$$

Clearly φ satisfies the assumptions of Theorem 1.15. But 0 is the only critical value of φ.

1.4   Semilinear Dirichlet problem

In this section, we consider the model problem

$$(\mathcal{P}_1) \qquad \begin{cases} -\Delta u + \lambda u = |u|^{p-2}u, \\ u \geq 0, u \in H_0^1(\Omega), \end{cases}$$

where Ω is a domain of $\mathbb{R}^N$. The main result is the following:

Theorem 1.19. *Assume that $|\Omega| < \infty$ and $2 < p < 2^*$. Then problem $(\mathcal{P}_1)$ has a nontrivial solution if and only if $\lambda > -\lambda_1(\Omega)$.*

Proof. **Necessary condition.** Suppose u is a nontrivial solution of $(\mathcal{P}_1)$. Let $e_1 \in H_0^1$ be an eigenfunction of $-\Delta$ corresponding to $\lambda_1 = \lambda_1(\Omega)$ with $e_1 > 0$ on Ω (see [90]). We have

$$\lambda \int_\Omega u \, e_1 = \int_\Omega (u^{p-1} + \Delta u)e_1 > \int_\Omega \Delta u \, e_1 = -\lambda_1 \int_\Omega u \, e_1$$

and thus $\lambda > -\lambda_1$.

Sufficient condition. Suppose $\lambda > -\lambda_1$, so that $c_1 := 1 + \min(0, \lambda/\lambda_1) > 0$. On H_0^1 we have, by the Poincaré inequality,

$$|\nabla u|_2^2 + \lambda |u|_2^2 \geq c_1 |\nabla u|_2^2.$$

On H_0^1 we choose the norm $\|u\| := \sqrt{|\nabla u|_2^2 + \lambda |u|_2^2}$. Let us define $f(u) := (u^+)^{p-1}$ and $F(u) := (u^+)^p/p$.

By Corollary 1.13, the functional

$$\varphi(u) := \int_\Omega \left[\frac{|\nabla u|^2}{2} + \lambda \frac{u^2}{2} - F(u) \right]$$

is of class $C^2(H_0^1, \mathbb{R})$. We will verify the assumptions of the mountain pass theorem. The $(PS)_c$ condition follows from the next lemma. By the Sobolev theorem, $c_2 > 0$ such that, on H_0^1,

$$|u|_p \leq c_2 \|u\|.$$

Hence we obtain

$$\begin{aligned} \varphi(u) &\geq \frac{1}{2}\|u\|^2 - \frac{1}{p}|u|_p^p \\ &\geq \frac{1}{2}\|u\|^2 - \frac{c_2^p}{p}\|u\|^p \end{aligned}$$

and there exists $r > 0$ such that

$$b := \inf_{\|u\|=r} \varphi(u) > 0 = \varphi(0).$$

Let $u \in H_0^1$ with $u > 0$ on Ω. We have, for $t \geq 0$,

$$\varphi(tu) = \frac{t^2}{2}(|\nabla u|_2^2 + \lambda|u|_2^2) - \frac{t^p}{p}|u|_p^p.$$

Since $p > 2$, there exists $e := tu$ such that $||e|| > r$ and $\varphi(e) \leq 0$.

By the mountain pass theorem, φ has a positive critical value and problem

$$-\Delta u + \lambda u = f(u),$$
$$u \in H_0^1(\Omega),$$

has a nontrivial solution u. Multiplying the equation by u^- and integrating over Ω, we find

$$0 = |\nabla u^-|_2^2 + \lambda|u^-|_2^2 = ||u^-||^2.$$

Hence $u^- = 0$ and u is a solution of $(\mathcal{P}_1)$. $\square$

Lemma 1.20. *Under the assumptions p of Theorem 1.19, if $\lambda > -\lambda_1$ any sequence $(u_n) \subset H_0^1$ such that*

$$d := \sup_n \varphi(u_n) < \infty, \varphi'(u_n) \to 0$$

contains a convergent subsequence.

Proof. 1) For n big enough, we have

$$\begin{aligned} d + 1 + ||u_n|| &\geq \varphi(u_n) - p^{-1}\langle\varphi'(u_n), u_n\rangle \\ &= (\frac{1}{2} - \frac{1}{p})(|\nabla u_n|_2^2 + \lambda|u_n|_2^2) \\ &= (\frac{1}{2} - \frac{1}{p})||u_n||^2. \end{aligned}$$

It follows that $||u_n||$ is bounded.

2) Going if necessary to a subsequence, we can assume that $u_n \rightharpoonup u$ in H_0^1. By the Rellich theorem, $u_n \to u$ in L^p. Theorem A.2 implies that $f(u_n) \to f(u)$ in L^q where $q := p/(p-1)$. Observe that

$$||u_n - u||^2 = \langle\varphi'(u_n) - \varphi'(u), u_n - u\rangle + \int_\Omega (f(u_n) - f(u)(u_n - u)).$$

It is clear that

$$\langle\varphi'(u_n) - \varphi'(u), u_n - u\rangle \to 0, n \to \infty.$$

It follows from the Hölder inequality that

$$\left|\int_\Omega (f(u_n) - f(u))(u_n - u)\right| \leq |f(u_n) - f(u)|_q|u_n - u|_p \to 0, n \to \infty.$$

Thus we have proved that $||u_n - u|| \to 0, n \to \infty$. $\square$

1.5   Symmetry and compactness

Symmetry plays a basic role in variational problems. For example, the imbedding $H^1(\mathbb{R}^N) \subset L^2(\mathbb{R}^N)$ is noncompact because of the action of translations. If Ω is bounded, the embedding $H_0^1(\Omega) \subset L^{2^*}(\Omega)$ is noncompact because of the action of dilations. When the problem is invariant by a group of orthogonal transformations, the situation is different. In some cases, it suffices to consider invariant functions in order to recover compactness. We will also see in chapter 3, that, in other cases, symmetry implies multiplicity.

We will use the following lemma.

Lemma 1.21. (P.L. Lions, 1984). Let $r > 0$ and $2 \leq q < 2^*$. If (u_n) is bounded in $H^1(\mathbb{R}^N)$ and if

$$\sup_{y \in \mathbb{R}^N} \int_{B(y,r)} |u_n|^q \to 0, n \to \infty,$$

then $u_n \to 0$ in $L^p(\mathbb{R}^N)$ for $2 < p < 2^*$.

Proof. We consider the case $N \geq 3$. Let $q < s < 2^*$ and $u \in H^1(\mathbb{R}^N)$. Hölder and Sobolev inequalities imply that

$$
\begin{aligned}
|u|_{L^s(B(y,r))} &\leq |u|_{L^q(B(y,r))}^{1-\lambda} |u|_{L^{2^*}(B(y,r))}^{\lambda} \\
&\leq c|u|_{L^q(B(y,r))}^{1-\lambda} \left[\int_{B(y,r)} (|u|^2 + |\triangledown u|^2) \right]^{\lambda/2}
\end{aligned}
$$

where $\lambda := \frac{s-q}{2^*-q} \frac{2^*}{s}$. Choosing $\lambda = 2/s$, we obtain

$$\int_{B(y,r)} |u|^s \leq c^s |u|_{L^q(B(y,r))}^{(1-\lambda)s} \int_{B(y,r)} (|u|^2 + |\triangledown u|^2).$$

Now, covering $\mathbb{R}^N$ by balls of radius r, in such a way that each point of $\mathbb{R}^N$ is contained in at most $N + 1$ balls, we find

$$\int_{\mathbb{R}^N} |u|^s \leq (N+1)c^s \sup_{y \in \mathbb{R}^N} [\int_{B(y,r)} |u|^q]^{(1-\lambda)s/q} \int_{\mathbb{R}^N} (|u|^2 + |\triangledown u|^2).$$

Under the assumption of the lemma, $u_n \to 0$ in $L^s(\mathbb{R}^N)$. Since $2 < s < 2^*$, $u_n \to 0$ in $L^p(\mathbb{R}^N)$ for $2 < p < 2^*$, by Sobolev and Hölder inequalities. □

Definition 1.22. Let G be a subgroup of $\mathbf{O}(N)$, $y \in \mathbb{R}^N$ and $r > 0$. We define

$$m(y, r, G) := \sup\{n \in \mathbb{N} : \exists g_1, \ldots, g_n \in G : j \neq k \Rightarrow B(g_j y, r) \cap B(g_k y, r) = \phi\}.$$

An open subset Ω of $\mathbb{R}^N$ is invariant if $g\Omega = \Omega$ for every $g \in G$. An invariant subset Ω of $\mathbb{R}^N$ is compatible with G if, for some $r > 0$,

$$\lim_{\substack{|y| \to \infty \\ \text{dist}(y,\Omega) \leq r}} m(y, r, G) = \infty.$$

Definition 1.23. Let G be a subgroup of $\mathbf{O}(N)$ and let Ω be an invariant open subset of $\mathbb{R}^N$. The action of G on $H_0^1(\Omega)$ is defined by

$$gu(x) := u(g^{-1}x).$$

The subspace of invariant functions is defined by

$$H_{0,G}^1(\Omega) := \{u \in H_0^1(\Omega) : gu = u, \forall g \in G\}.$$

The following theorem is the main result of this section:

Theorem 1.24. If Ω is compatible with G, the following embeddings are compact:

$$H_{0,G}^1(\Omega) \subset L^p(\Omega), 2 < p < 2^*.$$

Proof. Assume that $u_n \rightharpoonup 0$ in $H_{0,G}^1(\Omega)$. It is clear that, for every n,

$$\int_{B(y,r)} |u_n|^2 \leq \sup_n |u_n|_2^2 / m(y, r, G).$$

Let $\varepsilon > 0$. If Ω is compatible with G, there exists $R > 0$ such that, for every n,

$$\sup_{|y| \geq R} \int_{B(y,r)} |u_n|^2 \leq \varepsilon.$$

It follows from the Rellich theorem that

$$\int_{B(0,R+r)} |u_n|^2 \to 0, n \to \infty,$$

and so

$$\sup_{|y| \leq R} \int_{B(y,r)} |u_n|^2 \to 0, n \to \infty.$$

By the preceding lemma, $u_n \to 0$ in $L^p(\Omega)$ for $2 < p < 2^*$. $\square$

Corollary 1.25. (P.L. Lions, 1982). Let $N_j \geq 2$, $j = 1, \ldots, k$, $\sum_{j=1}^k N_j = N$ and

$$G := \mathbf{O}(N_1) \times \mathbf{O}(N_2) \times \ldots \times \mathbf{O}(N_k).$$

Then the following embeddings are compact:

$$H_G^1(\mathbb{R}^N) \subset L^p(\mathbb{R}^N), 2 < p < 2^*.$$

Proof. It is easy to verify that $\mathbb{R}^N$ is compatible with G. $\square$

Corollary 1.26. (Strauss, 1977). *Let $N \geq 2$. Then the following embeddings are compact:*

$$H^1_{\mathbf{O}(N)}(\mathbb{R}^N) \subset L^p(\mathbb{R}^N), 2 < p < 2^*.$$

Proof. It suffices to apply the preceding result. $\square$

1.6 Symmetric solitary waves

This section is devoted to the problem

$(\mathcal{P}_2)$
$$\begin{cases} -\Delta u + u = |u|^{p-2}u, \\ u \in H^1(\mathbb{R}^N), \end{cases}$$

where $N \geq 2$ and $2 < p < 2^*$.

We will apply the mountain pass theorem to the functional

$$\varphi(u) := \int_{\mathbb{R}^N} \left[\frac{|\nabla u|^2}{2} + \frac{u^2}{2} - F(u) \right]$$

where $F(u) := (u^+)^p/p$. In fact it suffices to find the critical points of φ restricted to a subspace of invariant functions.

Definition 1.27. *The action of a topological group G on a normed space X is a continuous map*

$$G \times X \to X : [g, u] \to gu$$

such that

$$1 \cdot u = u,$$
$$(gh)u = g(hu),$$
$$u \mapsto gu \text{ is linear.}$$

The action is isometric if

$$\|gu\| = \|u\|.$$

The space of invariant points is defined by

$$\text{Fix}(G) := \{u \in X : gu = u, \forall g \in G\}.$$

A set $A \subset X$ is invariant if $gA = A$ for every $g \in G$. A function $\varphi : X \to \mathbb{R}$ is invariant if $\varphi \circ g = \varphi$ for every $g \in G$. A map $f : X \to X$ is equivariant if $g \circ f = f \circ g$ for every $g \in G$.

Theorem 1.28. (Principle of symmetric criticality, Palais, 1979). *Assume that the action of the topological group G on the Hilbert space X is isometric. If $\varphi \in \mathcal{C}^1(X, \mathbb{R})$ is invariant and if u is a critical point of φ restricted to $\text{Fix}(G)$ then u is a critical point of φ.*

Proof. 1) Since φ is invariant, we have

$$\langle \varphi'(gu), v \rangle = \lim_{t \to 0} \frac{\varphi(u + tg^{-1}v) - \varphi(u)}{t}$$
$$= \langle \varphi'(u), g^{-1}v \rangle.$$

2) Since the action is isometric, we obtain

$$(\nabla\varphi(gu), v) = (\nabla\varphi(u), g^{-1}v) = (g\nabla\varphi(u), v)$$

and so $\nabla\varphi$ is equivariant.

3) Assume that u is a critical point of φ restricted to $\text{Fix}(G)$. It is clear that

$$g\nabla\varphi(u) = \nabla\varphi(gu) = \nabla\varphi(u)$$

and so $\nabla\varphi(u) \in \text{Fix}(G)$. Hence

$$\nabla\varphi(u) \in \text{Fix}(G) \cap \text{Fix}(G)^{\perp} = \{0\}. \qquad \Box$$

Theorem 1.29. (Strauss, 1977). *If $N \geq 2$ and $2 < p < 2^*$, there exists a radially symmetric, positive, classical solution of $(\mathcal{P}_2)$.*

Proof. 1) Consider the functional φ restricted to $X := H^1_{\mathbf{O}(N)}(\mathbb{R}^N)$. We shall verify the assumptions of the mountain pass theorem. As in the proof of Theorem 1.19, there exists $e \in X$ and $r > 0$ such that $||e||_1 > r$ and

$$b := \inf_{||u||_1 = r} \varphi(u) > 0 = \varphi(0) \geq \varphi(e).$$

2) It remains to prove the Palais-Smale condition. Consider a sequence $(u_n) \subset X$ such that

$$\sup_n \varphi(u_n) < \infty, \varphi'(u_n) \to 0 \quad \text{in} \quad X'.$$

As in the proof of Lemma 1.20, $||u_n||_1$ is bounded. Going if necessary to a subsequence, we can assume that $u_n \rightharpoonup u$ in X. By Corollary 1.26, $u_n \to u$ in L^p. As in the proof of Lemma 1.20, it follows that $||u_n - u||_1 \to 0$.

3) Using the mountain pass theorem, we obtain a nontrivial critical point u of φ restricted to X. By the principle of symmetric criticality, we have

$$-\Delta u + u = (u^+)^{p-2}u.$$

Multiplying the equation by u^- and integrating over $\mathbb{R}^N$, we find

$$0 = |\nabla u^-|_2^2 + |u^-|_2^2 = ||u^-||_1^2.$$

Hence $u^- = 0$ and u is a nonnegative solution of $(\mathcal{P}_2)$.

4) The next lemma implies that $u \in \mathcal{C}^2(\mathbb{R}^N)$. By the strong maximum principle u is positive. $\Box$

Lemma 1.30. *If u is a solution of $(\mathcal{P}_2)$ then $u \in C^2(\mathbb{R}^N)$.*

Proof. Since

$$-\Delta u = au$$

where $a := |u|^{p-2} - 1 \in L^{N/2}_{loc}(\mathbb{R}^N)$, the Brézis-Kato theorem implies that $u \in L^p_{loc}(\mathbb{R}^N)$ for all $1 \le p < \infty$. Thus $u \in W^{2,p}_{loc}(\mathbb{R}^N)$ for all $1 \le p < \infty$. By elliptic regularity theory, $u \in C^2(\mathbb{R}^N)$. □

The existence of a nonradial solution of $(\mathcal{P}_2)$ has been an open problem for some time.

Theorem 1.31. (Bartsch-Willem, 1993). *If $N = 4$ or $N \ge 6$ and $2 < p < 2^*$ then problem $(\mathcal{P}_2)$ has a nonradial solution.*

Proof. Let $2 \le m \le N/2$ be a fixed integer different from $(N-1)/2$. The action of

$$G := \mathbf{O}(m) \times \mathbf{O}(m) \times \mathbf{O}(N - 2m)$$

on $H^1(\mathbb{R}^N)$ is defined by

$$gu(x) := u(g^{-1}x).$$

By Corollary 1.25, the embedding $H^1_G(\mathbb{R}^N) \subset L^p(\mathbb{R}^N)$ is compact. Let τ be the involution defined on $\mathbb{R}^N = \mathbb{R}^m \oplus \mathbb{R}^m \oplus \mathbb{R}^{N-2m}$ by

$$\tau(x_1, x_2, x_3) := (x_2, x_1, x_3).$$

The action of $H := \{\text{id}, \tau\}$ on $H^1_G(\mathbb{R}^N)$ is defined by

$$\begin{aligned} hu(x) \quad &:= u(x), &\quad h &= \text{id}, \\ &:= -u(h^{-1}x), &\quad h &= \tau. \end{aligned}$$

It is clear that 0 is the only radial function of

$$X := \{u \in H^1_G(\mathbb{R}^N) : hu = u, \forall h \in H\}.$$

Moreover the embedding $X \subset L^p(\mathbb{R}^N)$ is compact. As in the proof of Theorem 1.29, we apply the mountain pass theorem. We obtain a nontrivial critical point u of φ restricted to X. By the principle of symmetric criticality, u is a nontrivial critical point of φ. □

1.7 Subcritical Sobolev inequalities

Let $N \ge 2$ and $2 < p < 2^*$. The Sobolev theorem implies that

$$S_p := \inf_{\substack{u \in H^1(\mathbb{R}^N) \\ |u|_p = 1}} \|u\|_1^2 > 0.$$

In order to prove that the infimum is achieved, we consider a minimizing sequence $(u_n) \subset H^1(\mathbb{R}^N)$:

(1.8) $$|u_n|_p = 1, \quad ||u_n||_1^2 \to S_p, \quad n \to \infty.$$

Going if necessary to a subsequence, we may assume $u_n \rightharpoonup u$ in $H^1(\mathbb{R}^N)$, so that

$$||u||_1^2 \le \varliminf ||u_n||_1^2 = S_p.$$

Thus u is a minimizer provided $|u|_p = 1$. But we know only that $|u|_p \le 1$. Indeed, for any $v \in H^1$ and $y \in \mathbb{R}^N$ the translated function

$$v^y(x) := v(x + y)$$

satisfies

$$||v^y||_1 = ||v||_1, \quad |v^y|_p = |v|_p.$$

Hence the problem is invariant by the noncompact group of translations. In order to overcome this difficulty, we will use the following result.

Lemma 1.32. (Brézis-Lieb Lemma, 1983). *Let Ω be an open subset of $\mathbb{R}^N$ and let $(u_n) \subset L^p(\Omega)$, $1 \le p < \infty$. If*
a) (u_n) is bounded in $L^p(\Omega)$,
b) $u_n \to u$ almost everywhere on Ω, then

$$\lim_{n \to \infty} (|u_n|_p^p - |u_n - u|_p^p) = |u|_p^p.$$

Proof. Fatou's Lemma yields

$$|u|_p \le \varliminf |u_n|_p < \infty.$$

Fix $\varepsilon > 0$. There exists $c(\varepsilon)$ such that, for all $a, b \in \mathbb{R}$,

$$\left| |a + b|^p - |a|^p \right| \le \varepsilon |a|^p + c(\varepsilon)|b|^p.$$

Hence we obtain

$$\begin{aligned} f_n^\varepsilon &:= \left(\left| |u_n|^p - |u_n - u|^p - |u|^p \right| - \varepsilon |u_n - u|^p \right)^+ \\ &\le (1 + c(\varepsilon))|u|^p. \end{aligned}$$

By the Lebesgue theorem, $\int_\Omega f_n^\varepsilon \to 0$, $n \to \infty$. Since

$$\left| |u_n|^p - |u_n - u|^p - |u|^p \right| \le f_n^\varepsilon + \varepsilon |u_n - u|^p,$$

we obtain

$$\varlimsup_{n \to \infty} \int_\Omega \left| |u_n|^p - |u_n - u|^p - |u|^p \right| \le c\varepsilon$$

where $c := \sup_n |u_n - u|_p^p < \infty$. Now let $\varepsilon \to 0$. $\square$

Remarks 1.33. a) The preceding lemma is a refinement of Fatou's Lemma.

b) Under the assumptions of the lemma, $u_n \rightharpoonup u$ weakly in $L^p(\Omega)$. However, weak convergence in $L^p(\Omega)$ is not sufficient to obtain the conclusion, except when $p = 2$.

c) In any Hilbert space

$$u_n \rightharpoonup u \Rightarrow \lim_{n \to \infty} (|u_n|^2 - |u_n - u|^2) = |u|^2.$$

Theorem 1.34. (P.L. Lions, 1984). *Let $(u_n) \subset H^1(\mathbb{R}^N)$ be a minimizing sequence satisfying (1.8). Then there exists a sequence $(y_n) \subset \mathbb{R}^N$ such that $u_n^{y_n}$ contains a convergent subsequence. In particular there exists a minimizer for S_p.*

Proof. Since $|u_n|_p = 1$, Lemma 1.21 implies that

$$\delta := \lim_{n \to \infty} \sup_{y \in \mathbb{R}^N} \int_{B(y,1)} |u_n|^2 > 0.$$

Going if necessary to a subsequence, we may assume the existence of $(y_n) \subset \mathbb{R}^N$ such that

$$\int_{B(y_n,1)} |u_n|^2 > \delta/2.$$

Let us define $v_n := u_n^{y_n}$. Hence $|v_n|_p = 1$, $\|v_n\|_1^2 \to S_p$ and

(1.9)
$$\int_{B(0,1)} |v_n|^2 > \delta/2.$$

Since (v_n) is bounded in $H^1(\mathbb{R}^N)$, we may assume, going if necessary to a subsequence

$$\begin{aligned}
v_n &\rightharpoonup v &&\text{in } H^1(\mathbb{R}^N), \\
v_n &\to v &&\text{in } L^2_{\text{loc}}(\mathbb{R}^N), \\
v_n &\to v &&\text{a.e. on } \mathbb{R}^N.
\end{aligned}$$

By the preceding lemma,

$$1 = |v|_p^p + \lim |w_n|_p^p,$$

where $w_n := v_n - v$. Hence we have

$$\begin{aligned}
S_p = \lim \|v_n\|_1^2 &= \|v\|_1^2 + \lim \|w_n\|_1^2 \\
&\geq S_p[(|v|_p^p)^{2/p} + (1 - |v|_p^p)^{2/p}].
\end{aligned}$$

Since, by (1.9), $v \neq 0$, we obtain $|v|_p^p = 1$, and so

$$\|v\|_1^2 = S_p = \lim \|v_n\|_1^2. \qquad \square$$

Theorem 1.35. *There exists a radially symmetric, positive, C^2 minimizer for S_p.*

Proof. 1) By the preceding theorem, there exists a minimizer $u \in H^1(\mathbb{R}^N)$ for S_p. By Theorem C.4, u is radially symmetric. Replacing u by $|u|$, we may also assume that u is non-negative.

2) It follows from Lagrange multiplier rule that, for some $\lambda > 0$, u is a solution of

$$-\Delta u + u = \lambda u^{p-1}.$$

By Lemma 1.30, $u \in \mathcal{C}^2(\mathbb{R}^N)$. The strong maximum principle implies that u is positive. $\square$

1.8 Non symmetric solitary waves

This section is devoted to the problem

$$(\mathcal{P}_3) \qquad \begin{cases} -\Delta u + u = Q(x)|u|^{p-2}u, \\ u \geq 0, u \in H^1(\mathbb{R}^N), \end{cases}$$

where $N \geq 2$, $2 < p < 2^*$ and $Q \in \mathcal{C}(\mathbb{R}^N)$ satisfies

$$(1.10) \qquad 1 = \lim_{|x| \to \infty} Q(x) = \inf_{x \in \mathbb{R}^N} Q(x).$$

By scaling, it is easy to replace 1 by any positive number. Let us define as before $f(u) := (u^+)^{p-1}$ and $F(u) := (u^+)^p/p$. By a variant of Corollary 1.13, the functional

$$\varphi(u) := \int_{\mathbb{R}^N} \Big[\frac{|\nabla u|^2}{2} + \frac{u^2}{2} - Q(x)F(u) \Big] dx$$

is of class $\mathcal{C}^2(H^1(\mathbb{R}^N), \mathbb{R})$. Let $v > 0$ be a minimizing function for S_p and let $(a_n) \subset \mathbb{R}^N$ be such that $|a_n| \to \infty$, $n \to \infty$. It is easy to verify that

$$\varphi'\Big(S_p^{\frac{1}{p-2}} v^{a_n}\Big) \to 0, \varphi\Big(S_p^{\frac{1}{p-2}} v^{a_n}\Big) \to \Big(\frac{1}{2} - \frac{1}{p}\Big)S_p^{\frac{p}{p-2}}, n \to \infty.$$

Hence condition $(PS)_c$ is not satisfied for $c = (\frac{1}{2} - \frac{1}{p})S_p^{\frac{p}{p-2}}$.

Lemma 1.36. *Under assumption (1.10), any sequence $(u_n) \subset H^1(\mathbb{R}^N)$ such that*

$$d := \sup_n \varphi(u_n) < c^* := \Big(\frac{1}{2} - \frac{1}{p}\Big)S_p^{\frac{p}{p-2}}, \varphi'(u_n) \to 0$$

contains a convergent subsequence.

Proof. 1) As in the proof of Lemma 1.20, $(||u_n||_1)$ is bounded. Going if necessary to a subsequence, we can assume that

$$
\begin{aligned}
u_n &\rightharpoonup u && \text{in } H^1(\mathbb{R}^N),\\
u_n &\to u && \text{in } L^p_{\text{loc}}(\mathbb{R}^N),\\
u_n &\to u && \text{a.e. on } \mathbb{R}^N.
\end{aligned}
$$

It follows that

$$
f(u_n) \to f(u) \quad \text{in } L^{p/(p-1)}_{\text{loc}}(\mathbb{R}^N),
$$

and so

$$
-\Delta u + u = Q(x)|u|^{p-2}u,
$$

(1.11)
$$
\varphi(u) = \frac{||u||_1^2}{2} - \int Q(x)F(u)dx = \left(\frac{1}{2} - \frac{1}{p}\right)||u||_1^2 \geq 0.
$$

2) We write $v_n := u_n - u$. The Brézis-Lieb Lemma leads to

$$
\begin{aligned}
\int Q(x)F(u_n)dx &= \int Q(x)F(u)dx + \int Q(x)F(v_n)dx + o(1)\\
&= \int Q(x)F(u)dx + \int \frac{(v_n^+)^p}{p}dx + o(1).
\end{aligned}
$$

Assuming $\varphi(u_n) \to c \leq d$, we obtain

(1.12)
$$
\varphi(u) + \frac{||v_n||_1^2}{2} - \int \frac{(v_n^+)^p}{p}dx \to c.
$$

Since $\langle \varphi'(u_n), u_n \rangle \to 0$, we also obtain

$$
\begin{aligned}
||v_n||_1^2 - \int (v_n^+)^p dx &\to p\int Q(x)F(u)dx - ||u||_1^2\\
&= -\langle \varphi'(u), u \rangle\\
&= 0.
\end{aligned}
$$

We may therefore assume that

$$
||v_n||_1^2 \to b, \quad \int (v_n^+)^p \to b.
$$

By the Sobolev inequality, we have

$$
||v_n||_1^2 \geq S_p|v_n^+|_p^2,
$$

and so $b \geq S_p b^{2/p}$. Either $b = 0$ or $b \geq S_p^{\frac{p}{p-2}}$. If $b = 0$, the proof is complete. Assume $b \geq S_p^{\frac{p}{p-2}}$. We obtain from (1.11) and (1.12)

$$
c^* = \left(\frac{1}{2} - \frac{1}{p}\right)S_p^{\frac{p}{p-2}} \leq \left(\frac{1}{2} - \frac{1}{p}\right)b \leq c \leq d < c^*,
$$

a contradiction. $\square$

Theorem 1.37. (Ding-Ni, 1986). *Under assumption* (1.10), *if* $N \geq 2$ *and* $2 < p < 2^*$, *problem* $(\mathcal{P}_3)$ *has a nontrivial solution.*

Proof. 1) It suffices to apply the mountain pass theorem with a value $c < c^*$. Let $v > 0$ be a minimizing function for S_p. If $Q \equiv 1$, the result follows from Theorem 1.29. We may assume that $Q \not\equiv 1$. Hence we obtain $\int Q(x)v^p dx > \int v^p dx$. It follows that

$$
\begin{aligned}
0 < \max_{t \geq 0} \varphi(tv) &= \max_{t \geq 0}\left(\frac{t^2}{2}\|v\|_1^2 - \frac{t^p}{p}\int Q(x)v^p dx\right) \\
&= \left(\frac{1}{2} - \frac{1}{p}\right)\left[\|v\|_1^2/\left(\int Q(x)v^p dx\right)^{\frac{2}{p}}\right]^{\frac{p}{p-2}} \\
&< \left(\frac{1}{2} - \frac{1}{p}\right)[\|v\|_1^2/|v|_p^2]^{\frac{p}{p-2}} \\
&= \left(\frac{1}{2} - \frac{1}{p}\right)S_p^{\frac{p}{p-2}} = c^*.
\end{aligned}
$$

2) Since

$$
\begin{aligned}
\varphi(u) &\geq \frac{\|u\|_1^2}{2} - \frac{M}{p}|u|_p^p \\
&\geq \frac{\|u\|_1^2}{2} - \frac{M}{pS_p^{p/2}}\|u\|_1^p,
\end{aligned}
$$

where $M := \max_{\mathbb{R}^N} Q$, there exists $r > 0$ such that

$$
b := \inf_{\|u\|_1 = r} \varphi(u) > 0 = \varphi(0).
$$

There exists $t_0 > 0$ such that $\|t_0 v\|_1 > r$ and $\varphi(t_0 v) < 0$. It follows from the preceding step that

$$
\max_{t \in [0,1]} \varphi(tt_0 v) < c^*.
$$

By the preceding lemma and the mountain pass theorem, φ has a critical value $c \in [b, c^*[$ and problem

$$
\begin{cases}
-\Delta u + u = Q(x)f(u), \\
u \in H^1(\mathbb{R}^N),
\end{cases}
$$

has a nontrivial solution u. Multiplying the equation by u^- and integrating, we find $u^- = 0$ and u is a solution of $(\mathcal{P}_3)$. $\square$

1.9   Critical Sobolev inequality

Let $N \geq 3$. The optimal constant in the Sobolev inequality is given by

$$S := \inf_{\substack{u \in \mathcal{D}^{1,2}(\mathbb{R}^N) \\ |u|_{2^*} = 1}} |\nabla u|_2^2 > 0.$$

In order to prove that the infimum is achieved, we consider a minimizing sequence $(u_n) \subset \mathcal{D}^{1,2}(\mathbb{R}^N)$:

$$(1.13) \qquad |u_n|_{2^*} = 1, |\nabla u_n|_2^2 \to S, n \to \infty.$$

Going if necessary to a subsequence, we may assume $u_n \rightharpoonup u$ in $\mathcal{D}^{1,2}(\mathbb{R}^N)$, so that

$$|\nabla u|_2^2 \leq \varliminf |\nabla u_n|_2^2 = S.$$

Thus u is a minimizer provided $|u|_{2^*} = 1$. But we know only that $|u|_{2^*} \leq 1$. Indeed, for any $v \in \mathcal{D}^{1,2}$, $y \in \mathbb{R}^N$ and $\lambda > 0$, the rescaled function

$$v^{y,\lambda}(x) := \lambda^{(N-2)/2} v(\lambda x + y)$$

satisfies

$$|\nabla v^{y,\lambda}|_2 = |\nabla v|_2, \quad |v^{y,\lambda}|_{2^*} = |v|_{2^*}.$$

Hence the problem is invariant by translations and dilations. In order to exclude noncompactness, we will use some results from measure theory (see [90]).

Definition 1.38. *Let Ω be an open subset of $\mathbb{R}^N$ and define*

$$\mathcal{K}(\Omega) := \{u \in \mathcal{C}(\Omega) : \text{ supp } u \text{ is a compact subset of } \Omega\},$$

$$\mathcal{BC}(\Omega) := \{u \in \mathcal{C}(\Omega) : |u|_\infty := \sup_{x \in \Omega} |u(x)| < \infty\}.$$

The space $\mathcal{C}_0(\Omega)$ is the closure of $\mathcal{K}(\Omega)$ in $\mathcal{BC}(\Omega)$ with respect to the uniform norm. A finite measure on Ω is a continuous linear functional on $\mathcal{C}_0(\Omega)$. The norm of the finite measure μ is defined by

$$||\mu|| := \sup_{\substack{u \in \mathcal{C}_0(\Omega) \\ |u|_\infty = 1}} |\langle \mu, u \rangle|.$$

We denote by $\mathcal{M}(\Omega)$ (resp. $\mathcal{M}^+(\Omega)$) the space of finite measures (resp. positive finite measures) on Ω. A sequence (μ_n) converges weakly to μ in $\mathcal{M}(\Omega)$, written

$$\mu_n \rightharpoonup \mu,$$

provided

$$\langle \mu_n, u \rangle \to \langle \mu, u \rangle, \forall u \in \mathcal{C}_0(\Omega).$$

Theorem 1.39. *a) Every bounded sequence of finite measures on Ω contains a weakly convergent subsequence.*
   b) If $\mu_n \rightharpoonup \mu$ in $\mathcal{M}(\Omega)$ then (μ_n) is bounded and

$$||\mu|| \leq \underline{\lim} \, ||\mu_n||.$$

c) If $\mu \in \mathcal{M}^+(\Omega)$ then

$$||\mu|| = \langle \mu, 1 \rangle = \sup_{\substack{u \in B\mathcal{C}(\Omega) \\ |u|_\infty = 1}} \langle \mu, u \rangle.$$

Following P.L. Lions [51] (inequality 1.15), Bianchi, Chabrowski, Szulkin (inequality 1.16) and Ben-Naoum, Troestler, Willem (equalities 1.17 and 1.18), we describe the lack of compactness of the injection $\mathcal{D}^{1,2}(\mathbb{R}^N) \subset L^{2^*}(\mathbb{R}^N)$.

Lemma 1.40. (Concentration-compactness lemma). *Let $(u_n) \subset \mathcal{D}^{1,2}(\mathbb{R}^N)$ be a sequence such that*

$$
\begin{aligned}
u_n &\rightharpoonup u & &\text{in } \mathcal{D}^{1,2}(\mathbb{R}^N), \\
|\nabla(u_n - u)|^2 &\rightharpoonup \mu & &\text{in } \mathcal{M}(\mathbb{R}^N), \\
|u_n - u|^{2^*} &\rightharpoonup \nu & &\text{in } \mathcal{M}(\mathbb{R}^N), \\
u_n &\to u & &\text{a.e. on } \mathbb{R}^N
\end{aligned}
$$

and define

$$(1.14) \qquad \mu_\infty := \lim_{R\to\infty} \overline{\lim_{n\to\infty}} \int_{|x|\geq R} |\nabla u_n|^2, \qquad \nu_\infty := \lim_{R\to\infty} \overline{\lim_{n\to\infty}} \int_{|x|>R} |u_n|^{2^*}.$$

Then it follows that

$$(1.15) \qquad ||\nu||^{2/2^*} \leq S^{-1}||\mu||,$$

$$(1.16) \qquad \nu_\infty^{2/2^*} \leq S^{-1}\mu_\infty,$$

$$(1.17) \qquad \overline{\lim_{n\to\infty}} \, |\nabla u_n|_2^2 = |\nabla u|_2^2 + ||\mu|| + \mu_\infty,$$

$$(1.18) \qquad \overline{\lim_{n\to\infty}} \, |u_n|_{2^*}^{2^*} = |u|_{2^*}^{2^*} + ||\nu|| + \nu_\infty.$$

Moreover, if $u = 0$ and $||\nu||^{2/2^*} = S^{-1}||\mu||$, then ν and μ are concentrated at a single point.

Proof. 1) Assume first $u = 0$. Choosing $h \in \mathcal{D}(\mathbb{R}^N)$, we infer from the Sobolev inequality that

$$\left(\int |hu_n|^{2^*} dx \right)^{2/2^*} \leq S^{-1} \int |\nabla(hu_n)|^2 dx.$$

Since $u_n \to 0$ in L^2_{loc}, we obtain

$$(1.19) \qquad \left(\int |h|^{2^*} d\nu \right)^{2/2^*} \leq S^{-1} \int |h|^2 d\mu.$$

Inequality (1.15) then follows.

2) For $R > 1$, let $\psi_R \in \mathcal{C}^1(\mathbb{R}^N)$ be such that $\psi_R(x) = 1$ for $|x| > R+1$, $\psi_R(x) = 0$ for $|x| < R$ and $0 \leq \psi_R(x) \leq 1$ on $\mathbb{R}^N$. By the Sobolev inequality, we have

$$\left(\int |\psi_R u_n|^{2^*} dx \right)^{2/2^*} \leq S^{-1} \int |\nabla(\psi_R u_n)|^2 dx.$$

Since $u_n \to 0$ in L^2_{loc}, we obtain

$$(1.20) \qquad \varlimsup_{n \to \infty} \left(\int |\psi_R u_n|^{2^*} dx \right)^{2/2^*} \leq S^{-1} \varlimsup_{n \to \infty} \int |\nabla u_n|^2 \psi_R^2 dx.$$

On the other hand, we have

$$\int_{|x|>R+1} |\nabla u_n|^2 dx \leq \int |\nabla u_n|^2 \psi_R^2 dx \leq \int_{|x|>R} |\nabla u_n|^2 dx$$

and

$$\int_{|x|>R+1} |u_n|^{2^*} dx \leq \int |u_n|^{2^*} \psi_R^2 dx \leq \int_{|x|>R} |u_n|^{2^*} dx.$$

We obtain from (1.14)

$$\mu_\infty = \lim_{R \to \infty} \varlimsup_{n \to \infty} \int |\nabla u_n|^2 \psi_R^2 dx, \quad \nu_\infty = \lim_{R \to \infty} \varlimsup_{n \to \infty} \int |u_n|^{2^*} \psi_R^2 dx.$$

Inequality (1.16) follows then from (1.20).

3) Assume moreover that $||\nu||^{2/2^*} = S^{-1}||\mu||$. The Hölder inequality and (1.19) imply that, for $h \in \mathcal{D}(\mathbb{R}^N)$,

$$\left(\int |h|^{2^*} d\nu \right)^{1/2^*} \leq S^{-1/2} ||\mu||^{1/N} \left(\int |h|^{2^*} d\mu \right)^{1/2^*}.$$

We deduce $\nu = S^{-2^*/2} ||\mu||^{2/N-2} \mu$. It follows from (1.19) that, for $h \in \mathcal{D}(\mathbb{R}^N)$,

$$\left(\int |h|^{2^*} d\nu \right)^{1/2^*} ||\nu||^{1/N} \leq \left(\int |h|^2 d\nu \right)^{1/2}$$

and so, for each open set Ω,

$$\nu(\Omega)^{1/2^*}\nu(\mathbb{R}^N)^{1/N} \leq \nu(\Omega)^{1/2}.$$

It follows that ν is concentrated at a single point.

4) Considering now the general case, we write $v_n := u_n - u$. Since

$$v_n \rightharpoonup 0, \quad \text{in } \mathcal{D}^{1,2}(\mathbb{R}^N),$$

we have

$$|\nabla u_n|^2 \rightharpoonup \mu + |\nabla u|^2, \quad \text{in } \mathcal{M}(\mathbb{R}^N).$$

According to the Brézis-Lieb Lemma, we have for every non negative $h \in \mathcal{K}(\mathbb{R}^N)$,

$$\int h|u|^{2^*} = \lim_{n\to\infty}\left(\int h|u_n|^{2^*} - \int h|v_n|^{2^*}\right).$$

Hence we obtain

$$|u_n|^{2^*} \rightharpoonup \nu + |u|^{2^*} \text{ in } \mathcal{M}(\mathbb{R}^N).$$

Inequality (1.15) follows from the corresponding inequality for (v_n).

5) Since

$$\overline{\lim_{n\to\infty}} \int_{|x|>R}|\nabla v_n|^2 = \overline{\lim_{n\to\infty}} \int_{|x|>R}|\nabla u_n|^2 - \int_{|x|>R}|\nabla u|^2,$$

we obtain

$$\lim_{R\to\infty}\overline{\lim_{n\to\infty}} \int_{|x|>R}|\nabla v_n|^2 = \mu_\infty.$$

By the Brézis-Lieb Lemma, we have

$$\int_{|x|>R}|u|^{2^*} = \lim_{n\to\infty}\left(\int_{|x|>R}|u_n|^{2^*} - \int_{|x|>R}|v_n|^{2^*}\right)$$

and so

$$\lim_{R\to\infty}\overline{\lim_{n\to\infty}} \int_{|x|\geq R}|v_n|^{2^*} = \nu_\infty.$$

Inequality (1.16) follows then from the corresponding inequality for (v_n).

6) For every $R > 1$, we have

$$\overline{\lim_{n\to\infty}} \int |\nabla u_n|^2 = \overline{\lim_{n\to\infty}}\left(\int \psi_R|\nabla u_n|^2 + \int(1-\psi_R)|\nabla u_n|^2\right)$$

$$= \overline{\lim_{n\to\infty}} \int \psi_R|\nabla u_n|^2 + \int(1-\psi_R)d\mu + \int(1-\psi_R)|\nabla u|^2.$$

When $R \to \infty$, we obtain, by Lebesgue theorem,

$$\overline{\lim_{n\to\infty}} \int |\nabla u_n|^2 = \mu_\infty + \int d\mu + \int |\nabla u|^2 = \mu_\infty + \|\mu\| + |\nabla u|_2^2.$$

The proof of (1.18) is similar. $\square$

Theorem 1.41. *(P.L. Lions, 1985). Let $(u_n) \subset \mathcal{D}^{1,2}(\mathbb{R}^N)$ be a minimizing sequence satisfying (1.13). Then there exists a sequence $(y_n, \lambda_n) \subset \mathbb{R}^N \times]0, \infty[$ such that $(u_n^{y_n, \lambda_n})$ contains a convergent subsequence. In particular there exists a minimizer for S.*

Proof. Define the Lévy concentration functions

$$Q_n(\lambda) := \sup_{y \in \mathbb{R}^N} \int_{B(y,\lambda)} |u_n|^{2^*}.$$

Since, for every n,

$$\lim_{\lambda \to 0^+} Q_n(\lambda) = 0, \quad \lim_{\lambda \to \infty} Q_n(\lambda) = 1,$$

there exists $\lambda_n > 0$ such that $Q_n(\lambda_n) = 1/2$. Moreover, there exists $y_n \in \mathbb{R}^N$ such that

$$\int_{B(y_n, \lambda_n)} |u_n|^{2^*} = Q_n(\lambda_n) = 1/2,$$

since

$$\lim_{|y| \to \infty} \int_{B(y, \lambda_n)} |u_n|^{2^*} = 0.$$

Let us define $v_n := u_n^{y_n, \lambda_n}$. Hence $|v_n|_{2^*} = 1$, $|\nabla v_n|_2^2 \to S$ and

(1.21)
$$\frac{1}{2} = \int_{B(0,1)} |v_n|^{2^*} = \sup_{y \in \mathbb{R}^N} \int_{B(y,1)} |v_n|^{2^*}.$$

Since (v_n) is bounded in $\mathcal{D}^{1,2}(\mathbb{R}^N)$, we may assume, going if necessary to a subsequence,

$$
\begin{aligned}
v_n &\rightharpoonup v &&\text{in } \mathcal{D}^{1,2}(\mathbb{R}^N),\\
|\nabla(v_n - v)|^2 &\rightharpoonup \mu &&\text{in } \mathcal{M}(\mathbb{R}^N),\\
|v_n - v|^{2^*} &\rightharpoonup \nu &&\text{in } \mathcal{M}(\mathbb{R}^N),\\
v_n &\to v &&\text{a.e. on } \mathbb{R}^N.
\end{aligned}
$$

By the preceding lemma,

(1.22)
$$S = \lim |\nabla v_n|_2^2 = |\nabla v|_2^2 + \|\mu\| + \mu_\infty,$$

(1.23)
$$1 = |v_n|_{2^*}^{2^*} = |v|_{2^*}^{2^*} + \|\nu\| + \nu_\infty,$$

where

$$\mu_\infty := \lim_{R \to \infty} \overline{\lim_{n \to \infty}} \int_{|x| > R} |\nabla v_n|^2, \quad \nu_\infty := \lim_{R \to \infty} \overline{\lim_{n \to \infty}} \int_{|x| > R} |v_n|^{2^*}.$$

We deduce from (1.22), (1.15), (1.16) and Sobolev inequality,

$$S \geq S\left((|v|_{2^*}^{2^*})^{2/2^*} + ||\nu||^{2/2^*} + \nu_\infty^{2/2^*}\right).$$

It follows from (1.23) that $|v|_{2^*}^{2^*}$, $||\nu||$ and ν_∞ are equal either to 0 or to 1. By (1.21), $\nu_\infty \leq 1/2$ so that $\nu_\infty = 0$. If $||\nu|| = 1$ then $v = 0$ and $||\nu||^{2/2^*} \geq S^{-1}||\mu||$. The preceding lemma implies that ν is concentrated at a single point z. We deduce from (1.21) the contradiction

$$\frac{1}{2} = \sup_{y \in \mathbb{R}^N} \int_{B(y,1)} |v_n|^{2^*} \geq \int_{B(z,1)} |v_n|^{2^*} \to ||\nu|| = 1.$$

Thus $|v|_{2^*}^{2^*} = 1$ and so

$$|\nabla v|_2^2 = S = \lim |\nabla v_n|_2^2. \qquad \square$$

Theorem 1.42. (Aubin, Talenti, 1976). *The instanton*

$$U(x) := \frac{[N(N-2)]^{(N-2)/4}}{[1 + |x|^2]^{(N-2)/2}}$$

is a minimizer for S.

Proof. 1) By the preceding theorem, there exists a minimizer $u \in \mathcal{D}^{1,2}(\mathbb{R}^N)$ for S. By Theorem C.4, u is radially symmetric. Replacing u by $|u|$, we may also assume that u is non-negative.

2) It follows from Lagrange multiplier rule that, for some $\lambda > 0$, u is a solution of

$$-\Delta u = \lambda u^{\frac{N+2}{N-2}}.$$

By the argument of Lemma 1.30, $u \in \mathcal{C}^2(\mathbb{R}^N)$. The strong maximum principle implies that u is positive.

3) After scaling, we may assume

$$-\Delta u = u^{\frac{N+2}{N-2}}.$$

Moreover we can choose $\varepsilon > 0$ such that

$$U_\varepsilon(x) := \varepsilon^{(2-N)/2} U(x/\varepsilon)$$

satisfies

$$U_\varepsilon(0) = u(0).$$

But then u and U_ε are solutions of the problem

$$\begin{cases} \partial_r(r^{N-1}\partial_r v) = r^{N-1}v^{\frac{N+2}{N-2}}, r > 0, \\ v(0) = u(0) \quad \partial_r v(0) = 0. \end{cases}$$

It follows easily that $u = U_\varepsilon$. By invariance, U is a minimizer for S. $\square$

Proposition 1.43. *For every open subset Ω of $\mathbb{R}^N$,*

$$S(\Omega) := \inf_{\substack{u \in \mathcal{D}_0^{1,2}(\Omega) \\ |u|_{2^*}=1}} |\nabla u|_2^2 = S$$

and $S(\Omega)$ is never achieved except when $\Omega = \mathbb{R}^N$.

Proof. 1) It is clear that $S \leq S(\Omega)$. Let $(u_n) \subset \mathcal{D}(\mathbb{R}^N)$ be a minimizing sequence for S. We can choose $y_n \subset \mathbb{R}^N$ and $\lambda_n > 0$ such that

$$u_n^{y_n, \lambda_n} \in \mathcal{D}(\Omega).$$

Hence we obtain $S(\Omega) \leq S$.

2) Assume that $\Omega \neq \mathbb{R}^N$ and $u \in \mathcal{D}_0^{1,2}(\Omega)$ is a minimizer for $S(\Omega)$. By the preceding step, u is also a minimizer for S. We may assume that $u \geq 0$, so that u is a solution of

$$-\Delta u = \lambda u^{\frac{N+2}{N-2}}.$$

By the strong maximum principle, $u > 0$ on $\mathbb{R}^N$. This is a contradiction, since $u \in \mathcal{D}_0^{1,2}(\Omega)$. $\square$

1.10   Critical nonlinearities

This section is devoted to the problem

$(\mathcal{P}_4)$
$$\begin{cases} -\Delta u + \lambda u = |u|^{2^*-2} u, \\ u \geq 0, u \in H_0^1(\Omega), \end{cases}$$

where Ω is a bounded domain of $\mathbb{R}^N$, $N \geq 3$ and $\lambda > -\lambda_1(\Omega)$.

Let us define as before $f(u) := (u^+)^{2^*-1}$ and $F(u) := (u^+)^{2^*}/2^*$. By Corollary 1.13, the functional

$$\varphi(u) := \int_\Omega \left[\frac{|\nabla u|^2}{2} + \lambda \frac{u^2}{2} - F(u) \right] dx$$

is of class $\mathcal{C}^2(H_0^1(\Omega), \mathbb{R})$. On $H_0^1(\Omega)$, we choose the norm $\|u\| := \sqrt{|\nabla u|_2^2 + \lambda |u|_2^2}$.

Lemma 1.44. *Any sequence $(u_n) \subset H_0^1(\Omega)$ such that*

$$d := \sup_n \varphi(u_n) < c^* := S^{N/2}/N, \varphi'(u_n) \to 0,$$

contains a convergent subsequence.

Proof. 1) As in the proof of Lemma 1.20, $(\|u_n\|)$ is bounded. Going if necessary to a subsequence, we can assume that

$$\begin{aligned} u_n &\rightharpoonup u \quad \text{in } H_0^1(\Omega), \\ u_n &\to u \quad \text{in } L^2(\Omega), \\ u_n &\to u \quad \text{a.e. on } \Omega. \end{aligned}$$

Since (u_n) is bounded in $L^{2^*}(\Omega)$, $(f(u_n))$ is bounded in $L^{2N/(N+2)}(\Omega)$ and so (see [90])

$$f(u_n) \rightharpoonup f(u) \quad \text{in} \quad L^{2N/(N+2)}(\Omega).$$

It follows that

$$-\Delta u + \lambda u = f(u)$$

and

$$(1.24) \qquad \varphi(u) = \frac{\|u\|^2}{2} - \int F(u) = (\frac{1}{2} - \frac{1}{2^*})|u^+|_{2^*}^{2^*} \geq 0.$$

2) We write $v_n := u_n - u$. The Brézis-Lieb Lemma leads to

$$\int F(u_n) = \int F(u) + \int F(v_n) + o(1).$$

Assuming $\varphi(u_n) \to c \leq d$, we obtain

$$(1.25) \qquad \varphi(u) + \frac{\|v_n\|^2}{2} - \int F(v_n) \to c.$$

Since $\langle \varphi'(u_n), u_n \rangle \to 0$, we obtain also

$$\begin{aligned} \|v_n\|^2 - 2^* \int F(v_n) &\to 2^* \int F(u) - \|u\|^2 \\ &= -\langle \varphi'(u), u \rangle \\ &= 0. \end{aligned}$$

We may therefore assume that

$$\|v_n\|^2 \to b, \quad 2^* \int F(v_n) \to b.$$

Since $v_n \to 0$ in $L^2(\Omega)$, it follows that $|\nabla v_n|_2^2 \to b$. By Sobolev inequality, we have

$$|\nabla v_n|_2^2 \geq S|v_n^+|_{2^*}^2.$$

and so $b \geq S\, b^{2/2^*}$. Either $b = 0$ or $b \geq S^{N/2}$. If $b = 0$, the proof is complete. Assume $b \geq S^{N/2}$. We obtain, from (1.24) and (1.25),

$$c^* = (\frac{1}{2} - \frac{1}{2^*})S^{N/2} \leq (\frac{1}{2} - \frac{1}{2^*})b \leq c \leq d < c^*,$$

a contradiction. $\square$

Theorem 1.45. (Brézis-Nirenberg, 1983). *Let Ω be a bounded domain of $\mathbb{R}^N$, $N \geq 4$. If $-\lambda_1(\Omega) < \lambda < 0$, then problem $(\mathcal{P}_4)$ has a nontrivial solution.*

Proof. 1) It suffices to apply the mountain pass theorem with a value $c < c^*$. By the next lemma, there exists a nonnegative $v \in H_0^1 \backslash \{0\}$ such that

$$||v||^2 / |v|_{2^*}^2 < S.$$

We obtain

$$
\begin{aligned}
0 < \max_{t \geq 0} \varphi(tv) &= \max_{t \geq 0}\left(\frac{t^2}{2}||v||^2 - \frac{t^{2^*}}{2^*}\int v^{2^*}\right) \\
&= (||v||^2/|v|_{2^*}^2)^{N/2}/N \\
&< S^{N/2}/N = c^*.
\end{aligned}
$$

2) Since

$$
\begin{aligned}
\varphi(u) &\geq \frac{||u||^2}{2} - \frac{1}{2^*}|u|_{2^*}^{2^*} \\
&\geq \frac{||u||^2}{2} - \frac{1}{2^* S^{2^*/2}}|\nabla u|_2^{2^*},
\end{aligned}
$$

there exists $r > 0$ such that

$$b := \inf_{||u|| = r} \varphi(u) > 0 = \varphi(0).$$

There exists also $t_0 > 0$ such that $||t_0 v|| > r$ and $\varphi(t_0 v) < 0$. It follows from the preceding step that

$$\max_{t \in [0,1]} \varphi(t t_0 v) < c^*.$$

By the preceding lemma and the mountain pass theorem, φ has a critical value $c \in [b, c^*[$ and problem

$$
\begin{cases}
-\Delta u + \lambda u = f(u), \\
u \in H_0^1(\Omega),
\end{cases}
$$

has a nontrivial solution u. Multiplying the equation by u^- and integrating, we find $u^- = 0$ and u is a solution of $(\mathcal{P}_4)$. $\square$

If U is the instanton, we have, for $\lambda < 0$,

$$\frac{||U||^2}{|U|_{2^*}^2} = \frac{|\nabla U|_2^2 + \lambda|U|_2^2}{|U|_{2^*}^2} < \frac{|\nabla U|_2^2}{|U|_{2^*}^2} = S.$$

Since $U \notin H_0^1(\Omega)$, it is necessary to "concentrate" U near a point of Ω after multiplication by a trunction function.

Lemma 1.46. *Under the assumption of Theorem 1.45, there exists a nonnegative $v \in H_0^1(\Omega) \backslash \{0\}$ such that*

$$||v||^2/|v|_{2^*}^2 < S.$$

Proof. We may assume that $0 \in \Omega$. Let $\psi \in \mathcal{D}(\Omega)$ be a nonnegative function such that $\psi \equiv 1$ on $B(0, \rho)$, $\rho > 0$, and define, for $\varepsilon > 0$,

$$U_\varepsilon(x) := \varepsilon^{(2-N)/2} U(x/\varepsilon),$$
$$u_\varepsilon(x) := \psi(x) U_\varepsilon(x).$$

It follows from Theorem 1.42 that

$$|\nabla U_\varepsilon|_2^2 = |U_\varepsilon|_{2^*}^{2^*} = S^{N/2}.$$

As $\varepsilon \to 0^+$, we have that

$$\int_\Omega |\nabla u_\varepsilon|^2 = \int_{\mathbb{R}^N} |\nabla U_\varepsilon|^2 + O(\varepsilon^{N-2}) = S^{N/2} + O(\varepsilon^{N-2}),$$
$$\int_\Omega |u_\varepsilon|^{2^*} = \int_{\mathbb{R}^N} |U_\varepsilon|^{2^*} + O(\varepsilon^N) = S^{N/2} + O(\varepsilon^N),$$
$$\int_\Omega |u_\varepsilon|^2 = \int_{B(0,\rho)} |U_\varepsilon|^2 + O(\varepsilon^{N-2})$$
$$\geq \int_{B(0,\varepsilon)} \frac{[N(N-2)\varepsilon^2]^{\frac{N-2}{2}}}{[2\varepsilon^2]^{N-2}} + \int_{\varepsilon < |x| < \rho} \frac{[N(N-2)\varepsilon^2]^{\frac{N-2}{2}}}{[2|x|^2]^{N-2}} + O(\varepsilon^{N-2})$$
$$= \begin{cases} d\varepsilon^2 |\ell n \varepsilon| + O(\varepsilon^2), & \text{if } N = 4, \\ d\varepsilon^2 + O(\varepsilon^{N-2}), & \text{if } N \geq 5, \end{cases}$$

where d is a positive constant. If $N = 4$, we obtain

$$\frac{||u_\varepsilon||^2}{|u|_{2^*}^2} \leq \frac{S^2 + \lambda d\varepsilon^2 |\ell n \varepsilon| + O(\varepsilon^2)}{(S^2 + O(\varepsilon^4))^{1/2}}$$
$$= S + \lambda d\varepsilon^2 |\ell n \varepsilon| S^{-1} + O(\varepsilon^2) < S,$$

for $\varepsilon > 0$ sufficiently small. And similarly, if $N \geq 5$, we have

$$\frac{||u_\varepsilon||^2}{|u_\varepsilon|_{2^*}^2} \leq \frac{S^{N/2} + \lambda d\varepsilon^2 + O(\varepsilon^{N-2})}{(S^{N/2} + O(\varepsilon^N))^{2/2^*}}$$
$$= S + \lambda d\varepsilon^2 S^{(2-N)/2} + O(\varepsilon^{N-2}) < S,$$

for $\varepsilon > 0$ sufficiently small. $\square$

When Ω is a smooth starshaped bounded domain, Theorem 1.45 is sharp.

Proposition 1.47. *Assume that problem* $(\mathcal{P}_4)$ *has a nontrivial solution. Then we have* $\lambda > -\lambda_1(\Omega)$. *Moreover if* Ω *is a smooth starshaped bounded domain, then* $\lambda < 0$.

Proof. As in Theorem 1.19, it is easy to see that $\lambda > -\lambda_1(\Omega)$. Let us prove that any nontrivial solution u of $(\mathcal{P}_4)$ is smooth if Ω is smooth. Since

$$-\Delta u = au$$

where $a := u^{2^*-2} - \lambda \in L^{N/2}(\Omega)$, Brézis-Kato theorem implies that $u \in L^p(\Omega)$ for all $1 \leq p < \infty$. Thus $u \in W^{2,p}(\Omega)$ for all $1 \leq p < \infty$. By elliptic regularity theory, $u \in C^2(\Omega) \cap C^1(\bar{\Omega})$. The Pohozaev identity (Theorem B.1) leads to

$$-\lambda \int_\Omega u^2 = \int_{\partial\Omega} \frac{|\nabla u|^2}{2} \sigma \cdot \nu \, d\sigma.$$

If Ω is starshaped about the origin, we have $s \cdot n > 0$ on $\partial\Omega$. It follows that $\lambda \leq 0$. If $\lambda = 0$, then $\nabla u = 0$ on $\partial\Omega$ and we obtain from $(\mathcal{P}_4)$

$$0 = -\int_\Omega \Delta u = \int_\Omega u^{2^*-1},$$

so that $u = 0$. □

Remarks 1.48. a) It is interesting to compare Propositions 1.43 and 1.47. Under the stronger assumption that the domain Ω is starshaped, Proposition 1.47 gives the stronger conclusion that equation

$$(1.26) \qquad\qquad -\Delta u = |u|^{2^*-2}u$$

has no positive solution in $H_0^1(\Omega)$.

b) For some domains Ω, equation (1.26) has a positive solution in $H_0^1(\Omega)$ (see [21]). By Proposition 1.43, it is not possible to construct this solution by minimization.

Chapter 2

Linking theorem

2.1 Quantitative deformation lemma

In order to extend the quantitative deformation lemma to continuously differentiable functions defined on a Banach space, we use the notion of pseudogradient defined by Palais in 1966.

Definition 2.1. *Let M be a metric space, X a normed space and $h : M \to X' \backslash \{0\}$ a continuous mapping. A pseudogradient vector field for h on M is a locally Lipschitz continuous vector field $g : M \to X$ such that, for every $u \in M$,*

$$||g(u)|| \leq 2||h(u)||$$
$$\langle h(u), g(u) \rangle \geq ||h(u)||^2.$$

Lemma 2.2. *Under the assumptions of the preceding definition, there exists a pseudogradient vector field for h on M.*

Proof. For every $v \in M$, there exists $x \in X$ such that $||x|| = 1$ and

$$\langle h(v), x \rangle > \frac{2}{3}||h(v)||.$$

Define $y := \frac{3}{2}||h(v)||x$ so that

$$||y|| < 2||h(v)||, \langle h(v), y \rangle > ||h(v)||^2.$$

Since h is continuous, there exists an open neighborhood N_v of v such that

(2.1)
$$||y|| \leq 2||h(u)||, \langle h(u), y \rangle \geq ||h(u)||^2,$$

for every $u \in N_v$. The family

$$\mathcal{N} := \{N_v : v \in M\}$$

is an open covering of M. Since M is metric, hence paracompact, there exists a locally finite open covering $\mathcal{M} := \{M_i : i \in I\}$ of M finer than $\mathcal{N}$. For each $i \in I$, there exists $v \in M$ such that $M_i \subset N_v$. Hence there exists $y = y_i$ such that (2.1) is satisfied for every $u \in M_i$. Define, on M,

$$\rho_i(u) \quad := \quad \text{dist}(u, X \backslash M_i\}$$

$$g(u) \quad := \quad \sum_{i \in I} \frac{\rho_i(u)}{\sum_{j \in I} \rho_j(u)} y_i.$$

It is easy to verify that g is a pseudogradient vector field for h on M. $\square$

The following lemma was proved by the author in 1983.

Lemma 2.3. Let X be a Banach space, $\varphi \in \mathcal{C}^1(X, \mathbb{R})$, $S \subset X$, $c \in \mathbb{R}$, $\varepsilon, \delta > 0$ such that

$$(2.2) \qquad (\forall u \in \varphi^{-1}([c - 2\varepsilon, c + 2\varepsilon]) \cap S_{2\delta}) : ||\varphi'(u)|| \geq 8\varepsilon/\delta.$$

Then there exists $\eta \in \mathcal{C}([0, 1] \times X, X)$ such that
(i) $\eta(t, u) = u$, if $t = 0$ or if $u \notin \varphi^{-1}([c - 2\varepsilon, c + 2\varepsilon]) \cap S_{2\delta}$,
(ii) $\eta(1, \varphi^{c+\varepsilon} \cap S) \subset \varphi^{c-\varepsilon}$,
(iii) $\eta(t, .)$ is an homeomorphism of X, $\forall t \in [0, 1]$,
(iv) $||\eta(t, u) - u|| \leq \delta$, $\forall u \in X$, $\forall t \in [0, 1]$,
(v) $\varphi(\eta(., u))$ is non increasing, $\forall u \in X$,
(vi) $\varphi(\eta(t, u)) < c$, $\forall u \in \varphi^c \cap S_\delta$, $\forall t \in]0, 1]$.

Proof. By the preceding lemma, there exists a pseudo-gradient vector field g for φ' on $M := \{u \in X : \varphi'(u) \neq 0\}$. Let us define

$$A \quad := \quad \varphi^{-1}([c - 2\varepsilon, c + 2\varepsilon]) \cap S_{2\delta},$$
$$B \quad := \quad \varphi^{-1}([c - \varepsilon, c + \varepsilon]) \cap S_\delta,$$
$$\psi(u) \quad := \quad \text{dist}(u, X \backslash A)(\text{dist}(u, X \backslash A) + \text{dist}(u, B))^{-1},$$

so that ψ is locally Lipschitz continuous, $\psi = 1$ on B and $\psi = 0$ on $X \backslash A$. Let us also define the locally Lipschitz continuous vector field

$$f(u) \quad := \quad -\psi(u)||g(u)||^{-2} g(u), \quad u \in A,$$
$$\quad := 0, \qquad\qquad\qquad\qquad u \in X \backslash A.$$

By Definition 2.1 and assumption (2.2), $||f(u)|| \leq \delta/8\varepsilon$ on X. For each $u \in X$, the Cauchy problem

$$\frac{d}{dt}\sigma(t, u) = f(\sigma(t, u))$$
$$\sigma(0, u) = u$$

has a unique solution $\sigma(., u)$ defined on $\mathbb{R}$. Moreover, σ is continuous on $\mathbb{R} \times X$ (see e.g. [78]). Let us define η on $[0, 1] \times X$ by $\eta(t, u) := \sigma(8\varepsilon t, u)$. It follows from Definition 2.1 and assumption (2.2) that, for $t \geq 0$,

$$(2.3) \qquad ||\sigma(t, u) - u|| = ||\int_0^t f(\sigma(\tau, u))d\tau||$$

$$\leq \int_0^t ||f(\sigma(\tau, u))||d\tau \leq \delta t/8\varepsilon$$

and

$$(2.4) \qquad \frac{d}{dt}\varphi(\sigma(t, u)) = \langle \varphi'(\sigma(t, u)), \frac{d}{dt}\sigma(t, u) \rangle$$

$$= \langle \varphi'(\sigma(t, u)), f(\sigma(t, u)) \rangle$$

$$\leq -\psi(\sigma(t, u))/4.$$

It is then easy to verify (i), (iii), (iv), (v) and (vi). Let $u \in \varphi^{c+\varepsilon} \cap S$. If there is $t \in [0, 8\varepsilon]$ such that $\varphi(\sigma(t, u)) < c - \varepsilon$, then $\varphi(\sigma(8\varepsilon, u)) < c - \varepsilon$ and (ii) is satisfied. If

$$\sigma(t, u) \in \varphi^{-1}([c - \varepsilon, c + \varepsilon]), \forall t \in [0, 8\varepsilon],$$

we obtain from (2.3) and (2.4)

$$\varphi(\sigma(8\varepsilon, u)) = \varphi(u) + \int_0^{8\varepsilon} \frac{d}{dt}\varphi(\sigma(t, u))dt$$

$$\leq \varphi(u) - \frac{1}{4}\int_0^{8\varepsilon} \psi(\sigma(t, u))dt$$

$$= c + \varepsilon - 2\varepsilon = c - \varepsilon,$$

and (ii) is also satisfied. $\square$

2.2 Ekeland variational principle

As a first application, we consider a particular case of Ekeland variational principle.

Theorem 2.4. (Ekeland, 1974). *Let X be a Banach space, $\varphi \in \mathcal{C}^1(X, \mathbb{R})$ bounded below, $v \in X$ and $\varepsilon, \delta > 0$. If*

$$\varphi(v) \leq \inf_X \varphi + \varepsilon$$

there exists $u \in X$ such that

$$\varphi(u) \leq \inf_X \varphi + 2\varepsilon, ||\varphi'(u)|| < 8\varepsilon/\delta, ||u - v|| \leq 2\delta.$$

Proof. We apply the preceding lemma with $S := \{v\}$ and $c := \inf_X \varphi$. Assume that for every $u \in \varphi^{-1}([c, c + 2\varepsilon]) \cap S_{2\delta}, ||\varphi'(u)|| \geq 8\varepsilon/\delta$. Then $\eta(1, v) \in \varphi^{c-\varepsilon}$, contradicting the definition of c. $\square$

Corollary 2.5. *Let $\varphi \in \mathcal{C}^1(X, \mathbb{R})$ be bounded below. If φ satisfies condition $(PS)_c$ with $c := \inf_X \varphi$ then every minimizing sequence for φ contains a converging subsequence. In particular, there exists a minimizer for φ.*

Proof. Let $(v_n) \subset X$ be a minimizing sequence for φ. We apply the preceding theorem with

$$\varepsilon_n = \max\{1/n, \varphi(v_n) - c\}, \delta_n = \sqrt{\varepsilon_n}.$$

There exists a sequence $(u_n) \subset X$ such that

$$\varphi(u_n) \to c, \quad \varphi'(u_n) \to 0, \quad ||u_n - v_n|| \to 0.$$

It suffices then to use $(PS)_c$ condition. $\square$

Theorem 2.6. (Brézis-Nirenberg, 1991). *Let $\varphi \in \mathcal{C}^1(X, \mathbb{R})$. If*

$$c := \lim_{||u|| \to \infty} \varphi(u) \in \mathbb{R},$$

then for every $\varepsilon, \delta > 0$, $R > 2\delta$, there exists $u \in X$ such that
a) $c - 2\varepsilon \leq \varphi(u) \leq c + 2\varepsilon$,
b) $||u|| > R - 2\delta$,
c) $||\varphi'(u)|| < 8\varepsilon/\delta$.

Proof. Suppose the thesis is false. We apply Lemma 2.3 with $S := X \backslash B(0, R)$. By the definition of c, $\varphi^{c+\varepsilon} \cap S$ is unbounded and $\varphi^{c-\varepsilon} \subset B(0, r)$ for $r > 0$ large enough. By Lemma 2.3, $\eta(1, \varphi^{c+\varepsilon} \cap S) \subset \varphi^{c-\varepsilon}$ and

$$\varphi^{c+\varepsilon} \cap S \subset (\varphi^{c-\varepsilon})_\delta \subset B(0, r + \delta),$$

a contradiction. $\square$

Corollary 2.7. (Shujie Li, 1986). *Let $\varphi \in \mathcal{C}^1(X, \mathbb{R})$ be bounded below. If every sequence $(u_n) \subset X$ such that*

$$\varphi(u_n) \to c, \varphi'(u_n) \to 0$$

is bounded, then

$$\varphi(u) \to \infty, ||u|| \to \infty.$$

Proof. If the thesis is false, then $c := \lim_{||u|| \to \infty} \varphi(u) \in \mathbb{R}$. By the preceding theorem, there exists a sequence $(u_n) \subset X$ such that

$$\varphi(u_n) \to c, \quad \varphi'(u_n) \to 0, \quad ||u_n|| \to \infty. \qquad \square$$

2.3 General minimax principle

In this section we prove a general minimax principle and we give some applications.

Theorem 2.8. *Let X be a Banach space. Let M_0 be a closed subspace of the metric space M and $\Gamma_0 \subset C(M_0, X)$. Define*

$$\Gamma := \{\gamma \in C(M, X) : \gamma\big|_{M_0} \in \Gamma_0\}.$$

If $\varphi \in C^1(X, \mathbb{R})$ satisfies

$$(2.5) \qquad \infty > c := \inf_{\gamma \in \Gamma} \sup_{u \in M} \varphi(\gamma(u)) > a := \sup_{\gamma_0 \in \Gamma_0} \sup_{u \in M_0} \varphi(\gamma_0(u))$$

then, for every $\varepsilon \in\,]0, (c-a)/2[$, $\delta > 0$ and $\gamma \in \Gamma$ such that

$$(2.6) \qquad \sup_M \varphi \circ \gamma \le c + \varepsilon,$$

there exists $u \in X$ such that
a) $c - 2\varepsilon \le \varphi(u) \le c + 2\varepsilon$,
b) $\mathrm{dist}(u, \gamma(M)) \le 2\delta$,
c) $\|\varphi'(u)\| \le 8\varepsilon/\delta$.

Proof. Suppose the thesis is false. We apply Lemma 2.3 with $S := \gamma(M)$. We assume that

$$(2.7) \qquad c - 2\varepsilon > a.$$

We define $\beta(u) := \eta(1, \gamma(u))$. For every $u \in M_0$, we obtain, from (2.7),

$$\beta(u) = \eta(1, \gamma_0(u)) = \gamma_0(u),$$

so that $\beta \in \Gamma$. It follows from (2.6) that

$$\sup_{u \in M} \varphi(\beta(u)) = \sup_{u \in M} \varphi\big(\eta(1, \gamma(u))\big) \le c - \varepsilon,$$

contradicting the definition of c. $\square$

Theorem 2.9. *Under assumption (2.5), there exists a sequence $(u_n) \subset X$ satisfying*

$$\varphi(u_n) \to c, \quad \varphi'(u_n) \to 0.$$

In particular, if φ satisfies $(PS)_c$ condition, then c is a critical value of φ.

We give three examples where condition (2.5) is satisfied.

Theorem 2.10. (Mountain pass theorem, Ambrosetti-Rabinowitz, 1973). *Let X be a Banach space, $\varphi \in \mathcal{C}^1(X, \mathbb{R})$, $e \in X$ and $r > 0$ be such that $\|e\| > r$ and*

$$b := \inf_{\|u\|=r} \varphi(u) > \varphi(0) \geq \varphi(e).$$

If φ satisfies the $(PS)_c$ condition with

$$c := \inf_{\gamma \in \Gamma} \max_{t \in [0,1]} \varphi(\gamma(t)),$$
$$\Gamma := \{\gamma \in \mathcal{C}([0,1], X) : \gamma(0) = 0, \gamma(1) = e\},$$

then c is a critical value of φ.

Proof. It suffices to apply the preceding theorem with $M = [0, 1]$, $M_0 = \{0, 1\}$, $\Gamma_0 = \{\gamma_0\}$, $\gamma_0(0) = 0$ and $\gamma_0(1) = e$. $\square$

Theorem 2.11. (Saddle-point theorem, Rabinowitz, 1978). *Let $X = Y \oplus Z$ be a Banach space with $\dim Y < \infty$. Define, for $\rho > 0$,*

$$M := \{u \in Y : \|u\| \leq \rho\}, \quad M_0 := \{u \in Y : \|u\| = \rho\}.$$

Let $\varphi \in \mathcal{C}^1(X, \mathbb{R})$ be such that

$$b := \inf_Z \varphi > a := \max_{M_0} \varphi.$$

If φ satisfies the $(PS)_c$ condition with

$$c := \inf_{\gamma \in \Gamma} \max_{u \in M} \varphi(\gamma(u)),$$
$$\Gamma := \{\gamma \in \mathcal{C}(M, X) : \gamma\big|_{M_0} = \mathrm{id}\},$$

then c is a critical value of φ.

Proof. In order to apply Theorem 2.9, we have only to verify that $c \geq b$. Let us prove that, for every $\gamma \in \Gamma$, $\gamma(M) \cap Z \neq \phi$. Denote by P the projection onto Y such that $PZ = \{0\}$. If $\gamma(M) \cap Z = \phi$, then the map

$$u \mapsto \rho P\gamma(u)/\|P\gamma(u)\|$$

is a retraction from the ball M onto its boundary M_0. This is impossible since $\dim Y < \infty$ (see Theorem D.11). Hence we obtain, for every $\gamma \in \Gamma$,

$$\max_M \varphi \circ \gamma \geq b := \inf_Z \varphi,$$

so that $c \geq b$. $\square$

Theorem 2.12. (Linking theorem, Rabinowitz, 1978). *Let $X = Y \oplus Z$ be a Banach space with $\dim Y < \infty$. Let $\rho > r > 0$ and let $z \in Z$ be such that $||z|| = r$. Define*

$$
\begin{aligned}
M &:= \{u = y + \lambda z : ||u|| \le \rho, \lambda \ge 0, y \in Y\}, \\
M_0 &:= \{u = y + \lambda z : y \in Y, ||u|| = \rho \text{ and } \lambda \ge 0 \text{ or } ||u|| \le \rho \text{ and } \lambda = 0\}, \\
N &:= \{u \in Z : ||u|| = r\}.
\end{aligned}
$$

Let $\varphi \in \mathcal{C}^1(X, \mathbb{R})$ be such that

$$
b := \inf_{N} \varphi > a := \max_{M_0} \varphi.
$$

If φ satisfies the $(PS)_c$ condition with

$$
\begin{aligned}
c &:= \inf_{\gamma \in \Gamma} \max_{u \in M} \varphi(\gamma(u)), \\
\Gamma &:= \{\gamma \in \mathcal{C}(M, X) : \gamma\big|_{M_0} = \mathrm{id}\},
\end{aligned}
$$

then c is a critical value of φ.

Proof. In order to apply Theorem 2.9, we have only to verify that $c \ge b$. Let us prove that, for every $\gamma \in \Gamma$, $\gamma(M) \cap N \ne \phi$. Denote by P the projection onto Y such that $PZ = \{0\}$ and by R a retraction from $Y \oplus \mathbb{R}z \backslash \{z\}$ to M_0. If $\gamma(M) \cap N = \phi$, then the map

$$
u \mapsto R(P\gamma(u) + ||(1 - P)\gamma(u)||r^{-1}z)
$$

is a retraction from M to M_0. This is impossible since M is homeomorphic to a finite dimensional ball (see Theorem D.11). Hence we obtain, for every $\gamma \in \Gamma$,

$$
\max_{M} \varphi \circ \gamma \ge b := \inf_{N} \varphi,
$$

so that $c \ge b$. $\square$

2.4 Semilinear Dirichlet problem

In this section, linking theorem is applied to the problem

$$
(\mathcal{P}_1) \qquad \begin{cases} -\Delta u + a(x)u = f(x, u), \\ u \in H_0^1(\Omega), \end{cases}
$$

where Ω is a domain of $\mathbb{R}^N$ and $a \in L^{N/2}(\Omega)$ if $N \ge 3$.

Lemma 2.13. *If $N \geq 3$ and $a \in L^{N/2}(\Omega)$, the functional*

$$\chi : \mathcal{D}_0^{1,2}(\Omega) \to \mathbb{R} : u \mapsto \int_\Omega a(x)u^2 dx$$

is weakly continuous.

Proof. The functional χ is well defined by the Sobolev and Hölder inequalities. Assume that $u_n \rightharpoonup u$ in $\mathcal{D}_0^{1,2}$ and consider an arbitrary subsequence (v_n) of (u_n). Since

$$v_n \to u \quad \text{in} \quad L^2_{\text{loc}},$$

going if necessary to a subsequence, we can assume that

$$v_n \to u \quad \text{a.e. on} \quad \Omega.$$

Since (v_n) is bounded in L^{2^*}, (v_n^2) is bounded in $L^{N/(N-2)}$. Hence $v_n^2 \rightharpoonup u^2$ in $L^{N/(N-2)}$ (see [90]) and so

$$\int a(x)v_n^2 dx \to \int a(x)u^2 dx.$$

We have thus proved that χ is weakly continuous. $\square$

Lemma 2.14. *If $|\Omega| < \infty$, $N \geq 3$ and $a \in L^{N/2}(\Omega)$, then*

$$\lambda_1 := \inf_{\substack{u \in H_0^1(\Omega) \\ |u|_2 = 1}} \int_\Omega (|\nabla u|^2 + a(x)u^2) dx > -\infty.$$

Proof. Consider a minimizing sequence $(u_n) \subset H_0^1$:

$$|\nabla u_n|_2 = 1, \frac{1 + \chi(u_n)}{|u_n|_2^2} \to \lambda_1.$$

Going if necessary to a subsequence, we may assume $u_n \rightharpoonup u$ in H_0^1. It follows from the Rellich theorem and from the preceding lemma that

$$|u_n|_2^2 \to |u|_2^2, \chi(u_n) \to \chi(u).$$

Since $\lambda_1 < +\infty$, $u \neq 0$. Hence we obtain

$$\lambda_1 \geq \frac{|\nabla u|_2^2 + \chi(u)}{|u|_2^2}. \qquad\qquad \square$$

Let

$$\lambda_1 < \lambda_2 \leq \ldots \leq \lambda_n \leq 0 < \lambda_{n+1} \leq \ldots$$

be the sequence of eigenvalues of

$$\begin{cases} -\Delta u + a(x)u = \lambda u, \\ u \in H_0^1(\Omega) \end{cases}$$

where each eigenvalue is repeated according to its multiplicity. Let $e_1, e_2, e_3, \ldots$ be the corresponding orthonormal eigenfunctions in $L^2(\Omega)$.

Lemma 2.15. *Under the assumptions of the preceding lemma, if*

$$Y := \text{span}(e_1, \ldots, e_n),$$
$$Z := \{u \in H_0^1(\Omega) : \int_\Omega uv = 0, v \in Y\},$$

then

$$\delta := \inf_{\substack{u \in Z \\ |\nabla u|_2 = 1}} \int_\Omega (|\nabla u|^2 + a(x)u^2) dx > 0.$$

Proof. By definition, on Z we have

$$\int (|\nabla u|^2 + au^2) \geq \lambda_{n+1} \int u^2.$$

Consider a minimizing sequence $(u_n) \subset Z$:

$$|\nabla u_n|_2 = 1, \quad 1 + \chi(u_n) \to \delta.$$

Going if necessary to a subsequence, we may assume $u_n \rightharpoonup u$ in H_0^1. By Lemma 2.13,

$$\delta = 1 + \chi(u) \geq \int |\nabla u|^2 + \chi(u) \geq \lambda_{n+1} \int u^2.$$

If $u = 0$, $\delta = 1$ and if $u \neq 0$, $\delta \geq \lambda_{n+1} \int u^2 > 0$. $\square$

We consider now the functional

$$\psi(u) := \int_\Omega F(x, u) dx,$$

where

$$F(x, u) := \int_0^u f(x, s) ds.$$

Lemma 2.16. *Assume that $|\Omega| < \infty$, $f \in C(\bar{\Omega} \times \mathbb{R})$, and*

$$|f(x, u)| \leq c(1 + |u|^{p-1})$$

with $1 < p < \infty$ if $N = 1, 2$ and $1 < p \leq 2^*$ if $N \geq 3$. Then the functional ψ is of class $C^1(H_0^1(\Omega), \mathbb{R})$ and

$$\langle \psi'(u), h \rangle = \int_\Omega f(x, u) h \, dx.$$

Proof. **Existence of the Gateaux derivative.** Let $u, h \in H_0^1$. Given $x \in \Omega$ and $0 < |t| < 1$, by the mean value theorem, there exists $\lambda \in]0, 1[$ such that

$$|F(x, u(x) + th(x)) - F(x, u(x))|/|t|$$
$$= |f(x, u(x) + \lambda th(x))h(x)|$$
$$\leq c(1 + (|u(x)| + |h(x)|)^{p-1})|h(x)|$$
$$\leq c(1 + 2^{p-1}(|u(x)|^{p-1} + |h(x)|^{p-1}))|h(x)|.$$

The Hölder inequality implies that

$$(1 + 2^{p-1}(|u(x)|^{p-1} + |h(x)|^{p-1}))|h(x)| \in L^1(\Omega).$$

It follows from the Lebesgue theorem that

$$\langle \psi'(u), h \rangle = \int_\Omega f(x, u) h \, dx.$$

Continuity of the Gateaux derivative. Assume that $u_n \to u$ in H_0^1. By Sobolev imbedding theorem, $u_n \to u$ in L^p. It follows from Theorem A.2 that $f(x, u_n) \to f(x, u)$ in L^q where $q := p/(p-1)$. We obtain, by the Hölder inequality,

$$
\begin{aligned}
|\langle \psi'(u_n) - \psi'(u), h \rangle| &\leq |f(x, u_n) - f(x, u)|_q |h|_p \\
&\leq c_p |f(x, u_n) - f(x, u)|_q \|h\|_1
\end{aligned}
$$

and so

$$\|\psi'(u_n) - \psi'(u)\| \leq c_p |f(x, u_n) - f(x, u)|_q \to 0, \quad n \to \infty. \qquad \square$$

We will prove that, under some restrictive conditions, the functional

$$\varphi(u) := \int_\Omega \left(\frac{|\nabla u|^2}{2} + \frac{a(x)u^2}{2} - F(x, u) \right) dx$$

satisfies the $(PS)_c$ condition for every $c \in \mathbb{R}$.

Lemma 2.17. Assume that $|\Omega| < \infty$ and

(f_1) $a \in L^{N/2}(\Omega)$ if $N \geq 3$, $a \in L^q(\Omega)$, $q > 1$, if $N = 2$ and $a \in L^1(\Omega)$ if $N = 1$, $f \in C(\bar{\Omega} \times \mathbb{R})$ and, for some $1 < p < 2^*$, $c > 0$,

$$|f(x, u)| \leq c(1 + |u|^{p-1}),$$

(f_2) there exists $\alpha > 2$ and $R > 0$ such that

$$|u| \geq R \Rightarrow 0 < \alpha F(x, u) \leq u f(x, u).$$

Then any sequence $(u_n) \subset H_0^1(\Omega)$ such that

$$d := \sup_n \varphi(u_n) < \infty, \varphi'(u_n) \to 0,$$

contains a converging subsequence.

Proof. 1) We consider the case $N \geq 3$. On H_0^1, we choose the norm $||u|| := |\nabla u|_2$. After integrating, we obtain from (f_2) the existence of $c_1 > 0$ such that

$$(2.8) \qquad c_1(|u|^\alpha - 1) \leq F(x, u).$$

Let $\beta \in]\alpha^{-1}, 2^{-1}[$. For n big enough, we have, for some $c_2, c_3 > 0$,

$$
\begin{aligned}
d + 1 + ||u_n|| &\geq \varphi(u_n) - \beta\langle\varphi'(u_n), u_n\rangle \\
&= \int [(\frac{1}{2} - \beta)(|\nabla u_n|^2 + au^2) + \beta f(x, u_n)u_n - F(x, u_n)]dx \\
&\geq (\frac{1}{2} - \beta)(\delta||z_n||^2 + \lambda_1|y_n|_2^2) + (\alpha\beta - 1)\int F(x, u_n)dx - c_2 \\
&\geq (\frac{1}{2} - \beta)(\delta||z_n||^2 + \lambda_1|y_n|_2^2) + c_1(\alpha\beta - 1)|u_n|_\alpha^\alpha - c_3,
\end{aligned}
$$

where, according to Lemma 2.15, $u_n = y_n + z_n$, $y_n \in Y$, $z_n \in Z$. It is then easy to verify that (u_n) is bounded in H_0^1 using the fact that $\dim Y$ is finite.

2) Going if necessary to a subsequence, we can assume that $u_n \rightharpoonup u$ in H_0^1. By the Rellich theorem, $u_n \to u$ in L^p. Theorem A.2 implies that $f(x, u_n) \to f(x, u)$ in L^q where $q := p/(p-1)$. Observe that

$$||u_n - u||^2 = \langle\varphi'(u_n) - \varphi'(u), u_n - u\rangle + \int [(f(x, u_n) - f(x, u))(u_n - u) - a(u_n - u)^2]dx.$$

It is clear that

$$\langle\varphi'(u_n) - \varphi'(u), u_n - u\rangle \to 0.$$

By Lemma 2.13,

$$\int a(u_n - u)^2 dx \to 0.$$

It follows from the Hölder inequality that

$$\left|\int (f(x, u_n) - f(x, u))(u_n - u)dx\right| \leq |f(x, u_n) - f(x, u)|_q |u_n - u|_p \to 0, \quad n \to \infty.$$

Thus we have proved that $||u_n - u|| \to 0$, $n \to \infty$. $\square$

Theorem 2.18. *Assume $|\Omega| < \infty$, $(f_1), (f_2)$ and*

(f_3) $f(x, u) = o(|u|)$, $|u| \to 0$, *uniformly on Ω,*

(f_4) $\lambda_n \frac{u^2}{2} \leq F(x, u)$,

then problem $(\mathcal{P}_1)$ has a nontrivial solution.

Proof. 1) We consider the case $N \geq 3$. We shall verify the assumptions of the linking theorem. The $(PS)_c$ condition follows from the preceding lemma. As before, we choose the norm $||u|| := |\nabla u|_2$.

2) Using (f_1) and (f_3), we obtain

$$(\forall \varepsilon > 0)\,(\exists c_\varepsilon > 0) : |F(x, u)| \leq \varepsilon |u|^2 + c_\varepsilon |u|^p.$$

We deduce from Lemma 2.15 that, on Z,

$$\varphi(u) \geq \frac{\delta}{2}||u||^2 - \int (\varepsilon |u|^2 + c_\varepsilon |u|^p)$$

$$= \frac{\delta}{2}||u||^2 - \varepsilon |u|_2^2 - c_\varepsilon |u|_p^p.$$

By Sobolev imbedding theorem, there exists $r > 0$ such that

$$b := \inf_{\substack{||u||=r \\ u \in Z}} \varphi(u) > 0.$$

3) By assumption (f_4), on Y, we have

$$\varphi(u) \leq \int [\lambda_n \frac{u^2}{2} - F(x, u)]dx \leq 0.$$

Define $z := re_{n+1}/||e_{n+1}||$. It follows from (2.8) that

$$\varphi(u) \leq \frac{||u||^2}{2} + |a|_{N/2}\frac{|u|_{2^*}^2}{2} - c_1|u|_\alpha^\alpha + c_1|\Omega|.$$

Since, on the finite dimensional space $Y \oplus \mathbb{R}z$, all norms are equivalent, we have

$$\varphi(u) \to -\infty, ||u|| \to \infty, u \in Y \oplus \mathbb{R}z.$$

Thus there exists $\rho > r$ such that

$$0 = \max_{M_0} \varphi,$$

where

$$M_0 := \{u := y + \lambda z : y \in Y, ||u|| = \rho \text{ and } \lambda \geq 0 \text{ or } ||u|| \leq \rho \text{ and } \lambda = 0\}.$$

4) If $\lambda_1 > 0$, it suffices to use the mountain pass theorem instead of the linking theorem. $\square$

Corollary 2.19. *Assume that $|\Omega| < \infty$ and $2 < p < 2^*$. Then, for every $\lambda \in \mathbb{R}$, problem*

$$\begin{cases} -\Delta u + \lambda u = |u|^{p-2}u, \\ u \in H_0^1(\Omega), \end{cases}$$

has a nontrivial solution.

2.5 Location theorem

Under some restrictive assumptions, it is possible to localize critical points.

Theorem 2.20. *Let N be a closed subset of the Banach space X. Let M_0 be a closed subset of the metric space M and $\Gamma_0 \subset \mathcal{C}(M_0, X)$. Define*

$$\Gamma := \{\gamma \in \mathcal{C}(M, X) : \gamma\big|_{M_0} \in \Gamma_0\}.$$

If $\varphi \in \mathcal{C}^1(X, \mathbb{R})$, $\bar{\varepsilon}, \bar{\delta} > 0$ are such that

$$(2.9) \qquad \text{dist}(N, \gamma(M_0) \cap \varphi^{-1}([c - \bar{\varepsilon}, c + \bar{\varepsilon}])) \geq \bar{\delta}, \quad \text{for every} \quad \gamma \in \Gamma,$$

$$(2.10) \qquad\qquad N \cap \gamma(M) \neq \phi \quad \text{for every} \quad \gamma \in \Gamma,$$

$$(2.11) \qquad -\infty < c := \inf_{\gamma \in \Gamma} \sup_{u \in M} \varphi(\gamma(u)) \leq \inf_N \varphi,$$

then, for every $\varepsilon \in {]}0, \bar{\varepsilon}/2[$, $\delta \in {]}0, \bar{\delta}/2[$ there exists $u \in X$ satisfying
a) $c - 2\varepsilon \leq \varphi(u) \leq c + 2\varepsilon$,
b) $\text{dist}(u, N) \leq 2\delta$,
c) $\|\varphi'(u)\| \leq 8\varepsilon/\delta$.

Proof. Suppose the thesis is false. We apply Lemma 2.3 with $S := N$, φ replaced by $\tilde{\varphi} := -\varphi$ and c replaced by $\tilde{c} := -c$. Let $\gamma \in \Gamma$ be such that

$$\sup_M \varphi \circ \gamma < c + \varepsilon$$

and define implicitly β on M by

$$\gamma(u) = \eta(1, \beta(u)).$$

Observe that, by assumption (2.9),

$$\text{dist}(N, \gamma(M_0) \cap \tilde{\varphi}^{-1}([\tilde{c} - 2\varepsilon, \tilde{c} + 2\varepsilon]))$$
$$\geq \text{dist}(N, \gamma(M_0) \cap \tilde{\varphi}^{-1}([\tilde{c} - \bar{\varepsilon}, \tilde{c} + \bar{\varepsilon}])) \geq \bar{\delta} > 2\delta.$$

Hence, for every $u \in M_0$,

$$\eta(1, \gamma_0(u)) = \gamma_0(u) = \eta(1, \beta(u))$$

and $\beta = \gamma_0$ on M_0. Since $\beta \in \Gamma$, assumption (2.10) implies that $\beta(v) \in N$ for some $v \in M$. It follows from assumption (2.11) that

$$\beta(v) \in N \subset \tilde{\varphi}^{\tilde{c}}.$$

Finally we obtain

$$\gamma(v) = \eta(1, \beta(v)) \in \tilde{\varphi}^{\tilde{c}-\varepsilon}$$

and

$$c + \varepsilon \leq \varphi(\gamma(v)) < c + \varepsilon,$$

which is a contradiction. $\square$

Definition 2.21. *Let N be a closed subset of a Banach space X, $\varphi \in C^1(X, \mathbb{R})$ and $c \in \mathbb{R}$. The function φ satisfies the $(PS)_{N,c}$ condition if any sequence $(u_n) \subset X$ such that*

$$\varphi(u_n) \to c, \quad \varphi'(u_n) \to 0, \quad \text{dist}(u_n, N) \to 0,$$

has a convergent subsequence.

Theorem 2.22. *Under the assumptions of Theorem 2.20, if φ satisfies the $(PS)_{N,c}$ condition, then there exists $u \in N$ such that $\varphi'(u) = 0$ and $\varphi(u) = c$.*

Proof. Theorem 2.20 implies the existence of a sequence $(u_n) \subset X$ satisfying

$$\varphi(u_n) \to c, \quad \varphi'(u_n) \to 0, \quad \text{dist}(u_n, N) \to 0.$$

It suffices then to use $(PS)_{N,c}$ condition. $\square$

2.6 Critical nonlinearities

This section is devoted to the problem

$$(\mathcal{P}_2) \qquad \begin{cases} -\Delta u + \lambda u = |u|^{2^* - 2} u, \\ u \in H_0^1(\Omega), \end{cases}$$

where Ω is a bounded domain of $\mathbb{R}^N$, $N \geq 3$, and $\lambda \in \mathbb{R}$. The energy is defined on $H_0^1(\Omega)$ by

$$\varphi(u) := \int_\Omega \left[\frac{|\nabla u|^2}{2} + \frac{\lambda u^2}{2} - \frac{|u|^{2^*}}{2^*} \right] dx.$$

On $H_0^1(\Omega)$ we choose the norm $||u|| := |\nabla u|_2$.

Let

$$0 < \lambda_1 < \lambda_2 \leq \ldots \leq \lambda_n \leq -\lambda < \lambda_{n+1} < \ldots$$

be the sequence of eigenvalues of $-\Delta$ on $H_0^1(\Omega)$ and let $e_1, e_2, e_3, \ldots$ be the corresponding orthonormal eigenfunctions in $L^2(\Omega)$. Define

$$
\begin{aligned}
Y &:= \text{span}(e_1, \ldots, e_n), \\
Z &:= \{u \in H_0^1 : \int_\Omega uv = 0, v \in Y\}, \\
\delta &:= \inf_{\substack{u \in Z \\ |\nabla u|_2 = 1}} \int_\Omega (|\nabla u|^2 + \lambda u^2) dx > 0.
\end{aligned}
$$

Lemma 2.23. *Any sequence $(u_n) \subset H_0^1(\Omega)$ such that*

$$d := \sup_n \varphi(u_n) < c^* := S^{N/2}/N, \varphi'(u_n) \to 0$$

contains a convergent subsequence.

Proof. 1) Let $\beta \in]2^{*-1}, 2^{-1}[$. For n large enough, we have that

$$
\begin{aligned}
d + 1 + ||u_n|| &\geq \varphi(u_n) - \beta \langle \varphi'(u_n), u_n \rangle \\
&= \int \left[(\frac{1}{2} - \beta)(|\nabla u_n|^2 + \lambda u_n^2) + (\beta - \frac{1}{2^*})|u_n|^{2^*} \right] dx \\
&\geq (\frac{1}{2} - \beta)(\delta ||z_n||^2 + (\lambda_1 + \lambda)|y_n|_2^2) + (\beta - \frac{1}{2^*})|u_n|_{2^*}^{2^*}
\end{aligned}
$$

where $u_n = y_n + z_n$, $y_n \in Y$, $z_n \in Z$. It is then easy to verify that (u_n) is bounded in H_0^1 using the fact that $\dim Y$ is finite.

2) Going if necessary to a subsequence, we can assume that

$$
\begin{aligned}
u_n &\rightharpoonup u \quad \text{in } H_0^1, \\
u_n &\to u \quad \text{in } L^2, \\
u_n &\to u \quad \text{a.e. on } \Omega.
\end{aligned}
$$

It follows then, as in Lemma 1.44, that $u_n \to u$ in H_0^1. $\square$

Theorem 2.24. (Capozzi-Fortunato-Palmieri, 1985). *Let Ω be a bounded domain of $\mathbb{R}^N$, $N \geq 4$. If $\lambda_n < -\lambda < \lambda_{n+1}$, problem $(\mathcal{P}_2)$ has a nontrivial solution.*

Proof. 1) It suffices to apply the linking theorem with a value $c < c^*$. By the next lemma, there exists $z \in Z \backslash \{0\}$ such that

$$
\max_{Y \oplus \mathbb{R}z} \varphi < c^*.
$$

2) On Z we have that

$$
\varphi(u) \geq \frac{\delta}{2} ||u||^2 - \frac{|u|_{2^*}^{2^*}}{2^*}.
$$

By the Sobolev inequality, there exists $r > 0$ such that

$$
b := \inf_{\substack{||u||=r \\ u \in Z}} \varphi(u) > 0.
$$

3) It is clear that $\varphi \leq 0$ on Y. Since $Y \oplus \mathbb{R}z$ is finite dimensional and

$$
\varphi(u) = \frac{||u||^2}{2} + \lambda \frac{|u|_2^2}{2} - \frac{|u|_{2^*}^{2^*}}{2},
$$

we have that

$$
\varphi(u) \to -\infty, ||u|| \to \infty, u \in Y \oplus \mathbb{R}z.
$$

Thus there exists $\rho > r$ such that

$$
0 = \max_{M_0} \varphi
$$

where

$$M_0 := \{u := y + \lambda z : y \in Y, ||u|| = \rho \text{ and } \lambda \geq 0 \text{ or } ||u|| \leq \rho \text{ and } \lambda = 0\}.$$

Since

$$c \leq \max_{Y \oplus \mathbb{R}z} \varphi < c^*,$$

the proof is complete. $\square$

Lemma 2.25. *Under the assumptions of Theorem 2.24, there exists $z \in Z\backslash\{0\}$ such that*

$$\max_{Y \oplus \mathbb{R}z} \varphi < c^*.$$

Proof. 1) For $\varepsilon > 0$, we define u_ε as in the proof of Lemma 1.46. As $\varepsilon \to 0^+$, we have that

(2.12)
$$\begin{aligned}
|u_\varepsilon|_{2^*}^{2^*} &= S^{N/2} + O(\varepsilon^N), \\
|u_\varepsilon|_{2^*-1}^{2^*-1} &= O(\varepsilon^{\frac{N-2}{2}}), \\
|u_\varepsilon|_1 &= O(\varepsilon^{\frac{N-2}{2}}).
\end{aligned}$$

We define

$$\begin{aligned}
z_\varepsilon &:= u_\varepsilon - \sum_{k=1}^n (\int_\Omega u_\varepsilon e_k) e_k, \\
V_\varepsilon &:= Y \oplus \mathbb{R}u_\varepsilon = Y \oplus \mathbb{R}z_\varepsilon.
\end{aligned}$$

Since, for every $u \neq 0$,

$$\max_{t \geq 0} \varphi(tu) = \frac{1}{N}\Big(\frac{||u||^2 + \lambda|u|_2^2}{|u|_{2^*}^2}\Big)^{N/2},$$

it suffices to prove that, for some $\varepsilon > 0$,

$$m_\varepsilon := \max_{\substack{u \in V_\varepsilon \\ |u|_{2^*} = 1}} (||u||^2 + \lambda|u|_2^2) < S.$$

Assume that $u := y + tu_\varepsilon = \tilde{y} + tz_\varepsilon$, $t \geq 0$, is such that $|u|_{2^*} = 1$ and $||u||^2 + \lambda|u|_2^2 = m_\varepsilon$. It is clear that $t > 0$ and

$$|\tilde{y}|_2 \leq |u|_2 \leq c_1|u|_{2^*} = c_1$$

and so $t \leq c_2$, $|y| \leq c_3$.
    We obtain, by convexity,

$$\begin{aligned}
1 = |u|_{2^*}^{2^*} &\geq |tu_\varepsilon|_{2^*}^{2^*} + 2^* \int_\Omega (tu_\varepsilon)^{2^*-1} y \, dx \\
&\geq |tu_\varepsilon|_{2^*}^{2^*} - c_4|u_\varepsilon|_{2^*-1}^{2^*-1}|y|_2.
\end{aligned}$$

It follows from (2.12) that

$$
\begin{aligned}
m_\varepsilon \;&\le\; (\lambda_n + \lambda)|y|_2^2 + \frac{\|u_\varepsilon\|^2 + \lambda|u_\varepsilon|_2^2}{|u_\varepsilon|_{2^*}^2}|tu_\varepsilon|_{2^*}^2 + c_5|u_\varepsilon|_1|y|_2 \\
&\le\; (\lambda_n + \lambda)|y|_2^2 + \frac{\|u_\varepsilon\|^2 + \lambda|u_\varepsilon|_2^2}{|u_\varepsilon|_{2^*}^2}(1 + c_4|u_\varepsilon|_{2^*-1}^{2^*-1}|y|_2)^{\frac{2}{2^*}} + c_5|u_\varepsilon|_1|y|_2 \\
&\le\; (\lambda_n + \lambda)|y|_2^2 + \frac{\|u_\varepsilon\|^2 + \lambda|u_\varepsilon|_2^2}{|u_\varepsilon|_{2^*}^2}(1 + O(\varepsilon^{\frac{N-2}{2}})|y|_2) + O(\varepsilon^{\frac{N-2}{2}})|y|_2.
\end{aligned}
$$

As in the proof of Lemma 1.46, we have, if $N \ge 5$,

$$
\frac{\|u_\varepsilon\|^2 + \lambda|u_\varepsilon|_2^2}{|u_\varepsilon|_{2^*}^2} = S + \lambda d\varepsilon^2 + O(\varepsilon^{N-2}),
$$

where $d > 0$. We deduce that

$$
\begin{aligned}
m_\varepsilon \;&\le\; (\lambda_n + \lambda)|y|_2^2 + S + \lambda d\varepsilon^2 + O(\varepsilon^{N-2}) + O(\varepsilon^{\frac{N-2}{2}})|y|_2 \\
&=\; S + \lambda d\varepsilon^2 + O(\varepsilon^{N-2}) \\
&<\; S
\end{aligned}
$$

for $\varepsilon > 0$ sufficiently small. If $N = 4$, the proof is similar. $\square$

Chapter 3

Fountain theorem

3.1 Equivariant deformation

When a functional is invariant, we use a more precise version of the quantitative deformation lemma.

Lemma 3.1. *Assume that the compact group G acts isometrically on the Banach space X. Let $\varphi \in C^1(X, \mathbb{R})$ be invariant and let $S \subset X$ be invariant. Assume that $c \in \mathbb{R}$, $\varepsilon, \delta > 0$ satisfy (2.2). Then there exists $\eta \in C([0,1] \times X, X)$ satisfying properties (i)-(vi) of Lemma 2.3 and*

(vii) $\qquad\qquad \eta(t, .)$ *is equivariant for every* $t \in [0, 1]$.

Proof. 1) By Lemma 2.2, there exists a pseudogradient vector field v for φ' on $M := \{u \in X : \varphi'(u) \neq 0\}$. Let us define the equivariant vector field on M

$$w(u) := \int_G g^{-1}v(gu)dg,$$

where dg denotes the normalized Haar measure on G.

2) By definition, we have

$$(3.1) \qquad \langle \varphi'(gu), h \rangle = \lim_{t \to 0} \frac{\varphi(u + tg^{-1}h) - \varphi(u)}{t}$$
$$= \langle \varphi'(u), g^{-1}h \rangle.$$

Since $h \to g^{-1}h$ is a surjective isometry, we obtain

$$(3.2) \qquad\qquad \|\varphi'(gu)\| = \|\varphi'(u)\|.$$

3) It follows from (3.1) and (3.2) that

$$\|w(u)\| \leq \int_G \|g^{-1}v(gu)\|dg$$
$$= \int_G \|v(gu)\|dg$$
$$\leq 2\int_G \|\varphi'(gu)\|dg = 2\|\varphi'(u)\|,$$

$$\begin{aligned}
\langle \varphi'(u), w(u) \rangle &= \int_G \langle \varphi'(u), g^{-1}v(gu) \rangle dg \\
&= \int_G \langle \varphi'(gu), v(gu) \rangle dg \\
&\geq \int_G \|\varphi'(gu)\|^2 dg = \|\varphi'(u)\|^2.
\end{aligned}$$

4) Finally let $u \in M$ and $K := \{gu : g \in G\} \in M$. Since v is locally Lipschitz continuous, there exists $\delta > 0$ such that v is Lipschitz continuous with constant c on K_δ. Now, if $u_1, u_2 \in B(u, \delta)$, we obtain

$$\begin{aligned}
\|w(u_1) - w(u_2)\| &\leq \int_G \|g^{-1}(v(gu_1) - v(gu_2))\| dg \\
&= \int_G \|v(gu_1) - v(gu_2)\| dg \\
&\leq c \int_G \|g(u_1 - u_2)\| dg = c\|u_1 - u_2\|
\end{aligned}$$

and w is locally Lipschitz continuous. $\square$

3.2   Fountain theorem

We will prove that, under some assumptions, an invariant functional has infinitely many critical values.

This result depends on the notion of admissible action introduced by Thomas Bartsch.

Definition 3.2. *Assume that the compact group G acts diagonally on V^k*

$$g(v_1, \ldots, v_k) := (gv_1, \ldots, gv_k)$$

where V is a finite dimensional space. The action of G is admissible if every continuous equivariant map $\partial U \to V^{k-1}$, where U is an open bounded invariant neighborhood of 0 in V^k, $k \geq 2$, has a zero.

Example 3.3. *The Borsuk-Ulam Theorem says that the antipodal action of $G := \mathbb{Z}/2$ on $V := \mathbb{R}$ is admissible (see Theorem D.17).*

We consider the following situation

(A_1) *The compact group G acts isometrically on the Banach space $X = \underset{j \in \mathbb{N}}{\oplus} X_j$, the spaces X_j are invariant and there exists a finite dimensional space V such that, for every $j \in \mathbb{N}$, $X_j \simeq V$ and the action of G on V is admissible.*

In this chapter, we will use the following notations:

$$Y_k := \oplus_{j=0}^{k} X_j, \quad Z_k := \overline{\oplus_{j=k}^{\infty} X_j},$$

$$B_k := \{u \in Y_k : ||u|| \le \rho_k\}, \quad N_k := \{u \in Z_k : ||u|| = r_k\}$$

where $\rho_k > r_k > 0$.

Lemma 3.4. (Intersection lemma). *Under assumption (A_1), if $\gamma \in \mathcal{C}(B_k, X)$ is equivariant and if $\gamma\big|_{\partial B_k} = $ id, then $\gamma(B_k) \cap N_k \ne \phi$.*

Proof. Define $U := \{u \in B_k : ||\gamma(u)|| < r_k\}$. Since $\rho_k > r_k$ and $\gamma(0) = 0$, U is an open bounded invariant neighborhood of 0 in $Y_k \simeq V^k$. Denote by P_k the projector onto Y_{k-1} such that $P_k Z_k = \{0\}$. The continuous equivariant map

$$\partial U \to Y_{k-1} \simeq V^{k-1} : u \mapsto P_k \gamma(u)$$

has a zero, since the action of G is admissible. It follows that $\gamma(B_k) \cap N_k \ne \phi$. $\square$

Theorem 3.5. *Under assumption (A_1), let $\varphi \in \mathcal{C}^1(X, \mathbb{R})$ be an invariant functional. Define, for $k \ge 2$,*

$$c_k := \inf_{\gamma \in \Gamma_k} \max_{u \in B_k} \varphi(\gamma(u)),$$

$$\Gamma_k := \{\gamma \in \mathcal{C}(B_k, X) : \gamma \text{ is equivariant and } \gamma\big|_{\partial B_k} = \text{id}\}.$$

If

$$b_k := \inf_{\substack{u \in Z_k \\ ||u|| = r_k}} \varphi(u) > a_k := \max_{\substack{u \in Y_k \\ ||u|| = \rho_k}} \varphi(u),$$

then $c_k \ge b_k$ and, for every $\varepsilon \in]0, (c_k - a_k)/2[$, $\delta > 0$ and $\gamma \in \Gamma_k$ such that

(3.3)
$$\max_{B_k} \varphi \circ \gamma \le c_k + \varepsilon,$$

there exists $u \in X$ such that
a) $c_k - 2\varepsilon \le \varphi(u) \le c_k + 2\varepsilon$,
b) $\text{dist}(u, \gamma(B_k)) \le 2\delta$,
c) $||\varphi'(u)|| \le 8\varepsilon/\delta$.

Proof. By the preceding lemma, $c_k \ge b_k$. Suppose the thesis is false. We apply Lemma 3.1 with $S := \gamma(B_k)$. We assume that

(3.4)
$$c_k - 2\varepsilon > a_k.$$

We define $\beta(u) := \eta(1, \gamma(u))$. For every $u \in \partial B_k$, we obtain from (3.4)

$$\beta(u) = \eta(1, \gamma(u)) = \eta(1, u) = u.$$

Since, by (vii), β is equivariant it follows that $\beta \in \Gamma_k$. We obtain from (3.3)

$$\max_{u \in B_k} \varphi(\beta(u)) = \max_{u \in B_k} \varphi\big(\eta(1, \gamma(u))\big) \le c_k - \varepsilon,$$

contradicting the definition of c_k. $\square$

Theorem 3.6. (Fountain theorem, Bartsch, 1992). *Under assumption (A_1), let $\varphi \in \mathcal{C}^1(X, \mathbb{R})$ be an invariant functional. If, for every $k \in \mathbb{N}$, there exists $\rho_k > r_k > 0$ such that*

(A_2) $a_k := \max\limits_{\substack{u \in Y_k \\ \|u\| = \rho_k}} \varphi(u) \le 0$

(A_3) $b_k := \inf\limits_{\substack{u \in Z_k \\ \|u\| = r_k}} \varphi(u) \to \infty,\ k \to \infty,$

(A_4) *φ satisfies the $(PS)_c$ condition for every $c > 0$,*

then φ has an unbounded sequence of critical values.

  Proof. For k large enough, $b_k > 0$. The preceding theorem implies then the existence of a sequence $(u_n) \subset X$ satisfying

$$\varphi(u_n) \to c_k, \quad \varphi'(u_n) \to 0.$$

It follows from (A4) that c_k is a critical value of φ. Since $c_k \ge b_k$ and $b_k \to \infty$, $k \to \infty$, the proof is complete. $\square$

3.3 Semilinear Dirichlet problem

In this section, the fountain theorem is applied to the problem

$(\mathcal{P}_1)$ $\qquad \begin{cases} -\Delta u = f(x, u), \\ u \in H_0^1(\Omega), \end{cases}$

where Ω is a domain of $\mathbb{R}^N$. We assume that $|\Omega| < \infty$. On $H_0^1(\Omega)$ we choose the norm $\|u\| := |\nabla u|_2$. We define the functional

$$\varphi(u) := \int_\Omega \Big(\frac{|\nabla u|^2}{2} - F(x, u)\Big) dx$$

where

$$F(x, u) := \int_0^u f(x, s) ds.$$

We choose an orthonormal basis (e_j) of $H_0^1(\Omega)$ and we define $X_j := \mathbb{R}e_j$. On $H_0^1(\Omega)$ we consider the antipodal action of $\mathbb{Z}/2$.

Theorem 3.7. *Assume that* $|\Omega| < \infty$ *and*

(f_1) $f \in C(\bar{\Omega} \times \mathbb{R})$ *and for some* $2 < p < 2^*$, $c > 0$,

$$|f(x, u)| \le c(1 + |u|^{p-1}),$$

(f_2) *there exists* $\alpha > 2$ *and* $R > 0$ *such that*

$$|u| \ge R \Rightarrow 0 < \alpha F(x, u) \le u f(x, u),$$

(f_3) $f(x, -u) = -f(x, u)$, $\forall x \in \Omega$, $\forall u \in \mathbb{R}$.

Then problem ($\mathcal{P}_1$) *has a sequence of solutions* (u_k) *such that* $\varphi(u_k) \to \infty$, $k \to \infty$.

Proof. 1) By Lemmas 2.16 and 2.17, φ is continuously differentiable and satisfies the (PS)$_c$ condition for every $c \in \mathbb{R}$.

2) After integrating, we obtain from (f_2) the existence of $c_1 > 0$ such that

$$c_1(|u|^\alpha - 1) \le F(x, u).$$

Hence, we have

$$\varphi(u) \le \frac{||u||^2}{2} - c_1|u|_\alpha^\alpha + c_1|\Omega|.$$

Since on the finite-dimensional space Y_k all norms are equivalent, relation (A_2) is satisfied for every $\rho_k > 0$ large enough.

3) After integrating, we obtain from (f_1) the existence of $c_2 > 0$ such that

$$|F(x, u)| \le c_2(1 + |u|^p).$$

Let us define

$$\beta_k := \sup_{\substack{u \in Z_k \\ ||u||=1}} |u|_p$$

so that, on Z_k, we have

$$\begin{aligned} \varphi(u) &\ge \frac{||u||^2}{2} - c_2|u|_p^p - c_2|\Omega| \\ &\ge \frac{||u||^2}{2} - c_2\beta_k^p||u||^p - c_2|\Omega|. \end{aligned}$$

Choosing $r_k := (c_2 p \beta_k^p)^{1/(2-p)}$, we obtain, if $u \in Z_k$ and $||u|| = r_k$,

$$\varphi(k) \ge (\frac{1}{2} - \frac{1}{p})(c_2 p \beta_k^p)^{2/(2-p)} - c_2|\Omega|.$$

Since, by the next lemma, $\beta_k \to 0$, $k \to \infty$, relation (A_3) is proved. It suffices then to use the fountain theorem with the antipodal action of $\mathbb{Z}/2$. $\square$

Lemma 3.8. *If $1 \leq p < 2^*$ then we have that*

$$\beta_k := \sup_{\substack{u \in Z_k \\ \|u\|=1}} |u|_p \to 0, \quad k \to \infty.$$

Proof. It is clear that $0 < \beta_{k+1} \leq \beta_k$, so that $\beta_k \to \beta \geq 0$, $k \to \infty$. For every $k \geq 0$, there exists $u_k \in Z_k$ such that $\|u_k\| = 1$ and $|u_k|_p > \beta_k/2$. By definition of Z_k, $u_k \rightharpoonup 0$ in H_0^1. The Sobolev imbedding theorem implies that $u_k \to 0$ in L^p. Thus we have proved that $\beta = 0$. $\square$

Corollary 3.9. *Assume that $|\Omega| < \infty$ and $2 < p < 2^*$. Then, for every $\lambda \in \mathbb{R}$, problem*

$$\begin{cases} -\Delta u + \lambda u = |u|^{p-2}u, \\ u \in H_0^1(\Omega), \end{cases}$$

has infinitely many solutions.

3.4   Multiple solitary waves

In this section, we consider the problem

$(\mathcal{P}_2)$
$$\begin{cases} -\Delta u + u = f(x, u), \\ u \in H^1(\mathbb{R}^N). \end{cases}$$

We define the functional

$$\psi(u) := \int_{\mathbb{R}^N} F(x, u)dx,$$

where

$$F(x, u) := \int_0^u f(x, s)ds.$$

Lemma 3.10. *Assume that $f \in \mathcal{C}(\mathbb{R}^N \times \mathbb{R})$ and*

$$|f(x, u)| \leq c(|u| + |u|^{p-1})$$

with $2 < p < \infty$ if $N = 1, 2$ and $2 < p \leq 2^*$ if $N \geq 3$. Then the function ψ is of class $\mathcal{C}^1(H^1(\mathbb{R}^N), \mathbb{R})$ and

$$\langle \psi'(u), h \rangle = \int_{\mathbb{R}^N} f(x, u)h\, dx.$$

Proof. **Existence of the Gateaux derivative.** Let $u, h \in H^1$. Given $x \in \mathbb{R}^N$ and $0 < |t| < 1$, by the mean value theorem, there exists $\lambda \in]0, 1[$ such that

$$|F(x, u(x) + th(x)) - F(x, u(x))|/|t|$$
$$= |f(x, u(x) + \lambda th(x))h(x)|$$
$$\leq c\big(|u(x)| + |h(x)| + 2^{p-1}(|u(x)|^{p-1} + |h(x)|^{p-1})\big)|h(x)|.$$

The Hölder inequality implies that

$$\Big(|u(x)| + |h(x)| + 2^{p-1}(|u(x)|^{p-1} + |h(x)|^{p-1})\Big)|h(x)| \in L^1.$$

It follows from the Lebesgue theorem that

$$\langle \psi'(u), h \rangle = \int f(x, u)h \, dx.$$

Continuity of the Gateaux derivative. Assume that $u_n \to u$ in H^1. By the Sobolev imbedding theorem, $u_n \to u$ in $L^2 \cap L^p$. It follows from Theorem A.4 that $f(x, u_n) \to f(x, u)$ in $L^2 + L^q$ where $q := p/(p-1)$. We obtain, by Hölder inequality,

$$
\begin{aligned}
|\langle \psi'(u_n) - \psi'(u), h \rangle| &\leq |f(x, u_n) - f(x, u)|_{2\vee q}|h|_{2\wedge p} \\
&\leq c_p|f(x, u_n) - f(x, u)|_{2\vee q}\|h\|_1
\end{aligned}
$$

and so

$$\|\psi'(u_n) - \psi'(u)\| \leq c_p|f(x, u_n) - f(x, u)|_{2\vee q} \to 0, n \to \infty. \qquad \square$$

In order to prove the Palais-Smale condition, we define the functional

$$\varphi(u) := \int_{\mathbb{R}^N} \Big(\frac{|\nabla u|^2}{2} + \frac{u^2}{2} - F(x, u)\Big)dx$$

on the space $X := H^1_{\mathbf{O}(N)}(\mathbb{R}^N)$ of radial functions.

Lemma 3.11. *Assume that $N \geq 2$ and*

(f_1') $f \in C(\mathbb{R}^N \times \mathbb{R})$ *and, for some $2 < p < 2^*$, $c > 0$,*

$$|f(x, u)| \leq c(|u| + |u|^{p-1}),$$

(f_2') *there exists $\alpha > 2$ such that, for every $x \in \mathbb{R}^N$ and $u \in \mathbb{R}$,*

$$\alpha F(x, u) \leq uf(x, u),$$

(f_3') $f(x, u) = o(|u|), |u| \to 0$, *uniformly on $\mathbb{R}^N$.*

Then any sequence $(u_n) \subset X$ such that

$$d := \sup_n \varphi(u_n) < \infty, \varphi'(u_n) \to 0,$$

contains a convergent subsequence.

Proof. 1) For n large enough, we have

$$
\begin{aligned}
d + 1 + \|u_n\|_1 &\geq \varphi(u_n) - \alpha^{-1}\langle \varphi'(u_n), u_n \rangle \\
&= (\frac{1}{2} - \frac{1}{\alpha})\|u_n\|_1^2 + \int_{\mathbb{R}^N} \left(\alpha^{-1} u_n f(x, u_n) - F(x, u_n) \right) dx \\
&\geq (\frac{1}{2} - \frac{1}{\alpha})\|u_n\|_1^2.
\end{aligned}
$$

Thus (u_n) is bounded in X.

2) Going if necessary to a subsequence, we can assume that $u_n \rightharpoonup u$ in X. By Corollary 1.26, $u_n \to u$ in L^p. Observe that

$$
\|u_n - u\|_1^2 = \langle \varphi'(u_n) - \varphi'(u), u_n - u \rangle + \int \left(f(x, u_n) - f(x, u) \right)(u_n - u) dx.
$$

It is clear that

$$
\langle \varphi'(u_n) - \varphi'(u), u_n - u \rangle \to 0.
$$

Using (f_1') and (f_3'), we obtain

$$
(\forall \varepsilon > 0)\,(\exists c_\varepsilon > 0) : |f(x, u)| \leq \varepsilon |u| + c_\varepsilon |u|^{p-1}
$$

and so

$$
\begin{aligned}
\int &\left(f(x, u_n) - f(x, u) \right)(u_n - u) dx \\
&\leq \int \left[\varepsilon[|u_n| + |u|] + c_\varepsilon[|u_n|^{p-1} + |u|^{p-1}] \right](u_n - u) dx \\
&\leq 4\varepsilon[|u_n|_2^2 + |u|_2^2] + c_\varepsilon[|u_n|_p^{p-1} + |u|_p^{p-1}]|u_n - u|_p \\
&\leq c(\varepsilon + c_\varepsilon|u_n - u|_p)
\end{aligned}
$$

where c is independent of ε and of n. Thus we have proved that

$$
\int \left(f(x, u_n) - f(x, u) \right)(u_n - u) dx \to 0
$$

and $\|u_n - u\|_1 \to 0$. $\square$

Theorem 3.12. (Strauss, 1977). *Assume $N \geq 2$ and (f_1'), (f_2'), (f_3') and*

(f_4') *there exists $R > 0$ such that*

$$
\inf_{\substack{x \in \mathbb{R}^N \\ |u| \geq R}} F(x, u) > 0,
$$

(f_5') $f(gx, u) = f(x, u), \forall g \in O(N), \forall x \in \mathbb{R}^N, \forall u \in \mathbb{R}$,

(f_6') $f(x, -u) = -f(x, u), \forall x \in \mathbb{R}^N, \forall u \in \mathbb{R}$.

Then problem $(\mathcal{P}_2)$ has a sequence of radial solutions (u_k) such that $\varphi(u_k) \to \infty$, $k \to \infty$.

Proof. 1) By assumption (f_5') and the principle of symmetric criticality, any critical point of φ is a solution of $(\mathcal{P}_2)$. Since, by assumption (f_6'), φ is invariant under the antipodal action of $\mathbb{Z}/2$, we apply the fountain theorem. It follows from the preceding lemmas that φ is continuously differentiable and satisfies the $(PS)_c$ condition for every $c \in \mathbb{R}$.

2) We choose an orthonormal basis (e_j) of X and we define $X_j := \mathbb{R}e_j$. After integrating, we obtain from (f_1'), (f_2') and (f_4') the existence of $c_1 > 0$ such that

$$c_1(|u|^\alpha - |u|^2) \leq F(x, u).$$

Hence we have

$$\varphi(u) \leq \frac{\|u\|_1^2}{2} - c_1|u|_\alpha^\alpha + c_1|u|_2^2.$$

Since, on the finite dimensional space Y_k all norms are equivalent, relation (A_2) is satisfied for every $\rho_k > 0$ large enough.

3) Let us define

$$\beta_k := \sup_{\substack{u \in Z_k \\ \|u\|_1 = 1}} |u|_p.$$

Using the proof of Lemma 3.8, we see that $\beta_k \to 0$, $k \to \infty$. After integrating, we obtain from (f_1') and (f_3') the existence of $c_2 > 0$ such that

$$|F(x, u)| \leq \frac{|u|^2}{4} + c_2|u|^p.$$

On Z_k, we have

$$\varphi(u) \geq \frac{\|u\|_1^2}{2} - \frac{|u|_2^2}{4} - c_2|u|_p^p$$

$$\geq \frac{\|u\|_1^2}{4} - c_2\beta_k^p\|u\|_1^p.$$

Choosing $r_k := (2c_2p\beta_k^p)^{1/(2-p)}$, we obtain, if $u \in Z_k$ and $\|u\| = r_k$,

$$\varphi(u) \geq (\frac{1}{4} - \frac{1}{2p})(2c_2p\beta_k^p)^{2/(2-p)}.$$

Since $\beta_k \to 0$, $k \to \infty$, relation (A_3) is proved. $\square$

We prove now the existence of nonradial solutions of $(\mathcal{P}_2)$ where $N = 4$ or $N \geq 6$.

Theorem 3.13. (Bartsch-Willem, 1993). *Under assumptions (f_1')-(f_6'), suppose $N = 4$ or $N \geq 6$, then problem $(\mathcal{P}_2)$ has a sequence of nonradial solutions (u_k) such that $\varphi(u_k) \to \infty$, $k \to \infty$.*

Proof. It suffices to use the subspace X of $H^1(\mathbb{R}^N)$ defined in the proof of Theorem 1.31 and to apply the fountain theorem and the principle of symmetric criticality. $\square$

Corollary 3.14. *Assume that $N \geq 2$ and $2 < q < p < 2^*$. Then, for every $\lambda \in \mathbb{R}$, problem*

$$\begin{cases} -\Delta u + u = \lambda|u|^{q-2}u + |u|^{p-2}u, \\ u \in H^1(\mathbb{R}^N), \end{cases}$$

has infinitely many radial solutions. Moreover if $N = 4$ or $N \geq 6$, the problem has infinitely many nonradial solutions.

Condition $p < 2^*$ is sharp.

Proposition 3.15. *If $N \geq 3$ and $p \geq 2^*$, then 0 is the only solution of*

$$\begin{cases} -\Delta u + u = |u|^{p-2}u, \\ u \in H^1(\mathbb{R}^N) \cap L^p(\mathbb{R}^N). \end{cases}$$

Proof. Corollary B.4 leads to

$$\begin{aligned} 0 &= \int_{\mathbb{R}^N} \left[\frac{N-2}{2}(|u|^p - |u|^2) - N\left(\frac{|u|^p}{p} - \frac{|u|^2}{2}\right) \right] dx \\ &= \int_{\mathbb{R}^N} \left[\left(\frac{N-2}{2} - \frac{N}{p}\right)|u|^p + |u|^2 \right] dx. \end{aligned}$$

If $p \geq 2^*$, then $u = 0$. $\square$

When $N = 1$, the nontrivial solution is unique up to translations.

Theorem 3.16. (Berestycki-Lions, 1983). *For $p > 2$, the problem*

$$\begin{cases} -u'' + u = |u|^{p-2}u, \\ \lim_{x \to \pm\infty} u(x) = 0, \sup_{x \in \mathbb{R}} u(x) > 0, \end{cases}$$

has a unique solution up to translations and this solution satisfies

a) $u(x) = u(-x)$,
b) $u(x) > 0$, $x \in \mathbb{R}$,
c) $u'(x) < 0$, $x > 0$.

Proof. 1) Let α be the positive zero of the potential

$$F(u) := \frac{|u|^p}{p} - \frac{u^2}{2}$$

and let u be the solution of the Cauchy problem

$$\begin{cases} -u'' = F'(u), \\ u(0) = \alpha, u'(0) = 0. \end{cases}$$

By uniqueness, $u(-x) = u(x)$. By conservation of energy, we obtain

$$\frac{u'(x)^2}{2} + F(u(x)) = 0.$$

In particular $u(x) \leq \alpha$ for $x \in \text{dom } u$. If there exists x_0 such that $u(x_0) = 0$, then $u'(x_0) = 0$ and, by uniqueness, $u \equiv 0$ which is impossible. Since $0 < u(x) \leq \alpha$, u is defined on $\mathbb{R}$. By conservation of energy, we have that

$$u'(x) = -\sqrt{-2F(u(x))} < 0, x > 0.$$

In particular

$$\ell := \lim_{x \to +\infty} u(x)$$

exists. Since

$$\lim_{x \to +\infty} u'(x) = -\sqrt{2F(\ell)},$$

it follows that $\ell = 0$.

2) Let v be another solution. After a translation, we can assume that $v(0) = \max_{x \in \mathbb{R}} v(x) > 0$. By conservation of energy, we obtain

$$\frac{v'(x)^2}{2} + F(v(x)) = F(v(0)).$$

Thus $\lim_{x \to +\infty} \frac{v'(x)^2}{2} = F(v(0))$ so that $F(v(0)) = 0$ and $v(0) = \alpha$. It follows that $u = v$ by uniqueness. $\square$

3.5 A dual theorem

We will prove a dual version of the fountain theorem by using the Galerkin method.

We assume that (A_1) is satisfied.

Definition 3.17. *Let $\varphi \in C^1(X, \mathbb{R})$ and $c \in \mathbb{R}$. The function φ satisfies the $(PS)_c^*$ condition (with respect to (Y_n)) if any sequence $(u_{n_j}) \subset X$ such that*

$$n_j \to \infty, u_{n_j} \in Y_{n_j}, \varphi(u_{n_j}) \to c, \varphi\big|'_{Y_{n_j}}(u_{n_j}) \to 0$$

contains a subsequence converging to a critical point of φ.

Theorem 3.18. (Dual fountain theorem, Bartsch-Willem, 1995). *Under assumption (A_1), let $\varphi \in C^1(X, \mathbb{R})$ be an invariant functional. If, for every $k \geq k_0$, there exists $\rho_k > r_k > 0$ such that*

(B_1) $a_k := \displaystyle\inf_{\substack{u \in Z_k \\ \|u\| = \rho_k}} \varphi(u) \geq 0,$

(B_2) $b_k := \max\limits_{\substack{u \in Y_k \\ ||u||=r_k}} \varphi(u) < 0,$

(B_3) $d_k := \inf\limits_{\substack{u \in Z_k \\ ||u|| \leq \rho_k}} \varphi(u) \to 0, k \to \infty,$

(B_4) φ satisfies the $(PS)^*_c$ condition for every $c \in [d_{k_0}, 0[,$

then φ has a sequence of negative critical values converging to 0.

Proof. We fix $n \geq k \geq k_0$ and we define

$$
\begin{aligned}
Z^n_k &:= \oplus^n_{j=k} X_j, \\
B^n_k &:= \{u \in Z^n_k : ||u|| \leq \rho_k\}, \\
\Gamma^n_k &:= \{\gamma \in \mathcal{C}(B^n_k, Y_n) : \gamma \text{ is equivariant and } \gamma\big|_{\partial B^n_k} = \text{id}\}, \\
c^n_k &:= \sup\limits_{\gamma \in \Gamma^n_k} \min\limits_{u \in B^n_k} \varphi(\gamma(u)).
\end{aligned}
$$

By Theorem 3.5, applied to the functional $-\varphi$ defined on the space Y_n, $c^n_k \leq b_k$ and there exists $u_n \in Y_n$ such that

$$
c^n_k - 2/n \leq \varphi(u_n) \leq c^n_k + 2/n, ||\varphi\big|'_{X_n}(u_n)|| \leq 8/n.
$$

We obtain, by definition, that $d_k \leq c^n_k$. Using (B_4), we see that $(c^n_k)_{n \geq k}$ converges along a subsequence to a critical value $c_k \in [d_k, b_k]$ of φ as $n \to \infty$. It follows from (B_3) that $c_k \to 0$ as $k \to \infty$. $\square$

Remarks 3.19. a) The $(PS)^*_c$ condition implies the $(PS)_c$ condition. Assume that $(u_j) \subset X$ is such that

$$
\varphi(u_j) \to c, \quad \varphi'(u_j) \to 0.
$$

There exists sequences (v_{n_j}), (n_j) such that

$$
n_j \to \infty, v_{n_j} \in Y_{n_j}, v_{n_j} - u_j \to 0, \varphi(v_{n_j}) - \varphi(u_j) \to 0, \varphi'(v_{n_j}) - \varphi'(u_j) \to 0.
$$

By $(PS)^*_c$, the sequence (v_{n_j}) contains a convergent subsequence and so (u_j) contains also a convergent subsequence.

b) It is necessary to use the Galerkin method in the preceding proof because the intersection lemma is not valid in the infinite dimension. A similar situation occurs if we try to generalize the linking theorem or the saddle-point theorem when the dimension of Y is infinite. See [11], [36] and [48].

3.6 Concave and convex nonlinearities

In this section, we consider the model problem

$(\mathcal{P}_3)$
$$\begin{cases} -\Delta u = \mu |u|^{q-2}u + \lambda |u|^{p-2}u, \\ u \in H_0^1(\Omega), \end{cases}$$

where Ω is a domain of $\mathbb{R}^N$ and $1 < q < 2 < p < 2^*$. The combined effect of concave and convex nonlinearities was studied by Ambrosetti, Brézis and Cerami. They proved the existence of infinitely many solutions with negative energy for $0 < \mu << \lambda = 1$. The energy is defined on $H_0^1(\Omega)$ by

$$\varphi_{\lambda,\mu}(u) := \int_\Omega \left[\frac{|\nabla u|^2}{2} - \frac{\mu |u|^q}{q} - \frac{\lambda |u|^p}{p} \right] dx.$$

On $H_0^1(\Omega)$ we choose the norm $||u|| := |\nabla u|_2$.

Theorem 3.20. (Bartsch-Willem, 1995). *Assume that $|\Omega| < \infty$ and $1 < q < 2 < p < 2^*$.*

a) For every $\lambda > 0$, $\mu \in \mathbb{R}$, problem $(\mathcal{P}_3)$ has a sequence of solutions (u_k) such that $\varphi_{\lambda,\mu}(u_k) \to \infty$, $k \to \infty$.

b) For every $\mu > 0$, $\lambda \in \mathbb{R}$, problem $(\mathcal{P}_3)$ has a sequence of solutions (v_k) such that $\varphi_{\lambda,\mu}(v_k) < 0$ and $\varphi_{\lambda,\mu}(v_k) \to 0$, $k \to \infty$.

Proof. 1) Part a) is a corollary of Theorem 3.7. Part b) follows from Theorem 3.18. We choose an orthonormal basis (e_j) of H_0^1 and we define $X_j := \mathbb{R}e_j$. On H_0^1 we consider the antipodal action of $\mathbb{Z}/2$. We assume that $\mu > 0$ and we set $\varphi := \varphi_{\lambda,\mu}$.

2) In order to verify (B_1), we define

$$\beta_k := \sup_{\substack{u \in Z_k \\ ||u||=1}} |u|_q.$$

There exists $R > 0$ such that

$$||u|| \leq R \Rightarrow |\lambda| \frac{c_1}{p} ||u||^p \leq \frac{1}{4} ||u||^2$$

where

$$c_1 := \sup_{\substack{u \in H_0^1 \\ ||u||=1}} |u|_p.$$

Hence we obtain, for $u \in Z_k$, $||u|| \leq R$,

(3.5)
$$\varphi(u) \geq \frac{||u||^2}{2} - \mu \beta_k^q \frac{||u||^q}{q} - |\lambda| c_1 \frac{||u||^p}{p}$$
$$\geq \frac{||u||^2}{4} - \mu \beta_k^q \frac{||u||^q}{q}.$$

We choose $\rho_k := (4\mu\beta_k^q/q)^{1/(2-q)}$. Since, by Lemma 3.8, $\beta_k \to 0$, $k \to \infty$, it follows that $\rho_k \to 0$, $k \to \infty$. There exists k_0 such that $\rho_k \le R$ when $k \ge k_0$. Thus, for $k \ge k_0$, $u \in Z_k$ and $||u|| = \rho_k$, we have $\varphi(u) \ge 0$ and (B_1) is proved.

3) Since on the finite dimensional space Y_k all norms are equivalent, relation (B_2) is satisfied for every $r_k > 0$ small enough, when $\mu > 0$.

4) We obtain from (3.5), for $k \ge k_0$, $u \in Z_k$, $||u|| \le \rho_k$,

$$\varphi(u) \ge -\mu\beta_k^q \frac{||u||^q}{q} \ge -\mu\beta_k^q \frac{\rho_k^q}{q}.$$

Since $\beta_k \to 0$ and $\rho_k \to 0$, $k \to \infty$, relation (B_3) is also satisfied.

5) Finally we prove the $(PS)_c^*$ condition. Consider a sequence $(u_{n_j}) \subset H_0^1$ such that

$$n_j \to \infty, u_{n_j} \in Y_{n_j}, \varphi(u_{n_j}) \to c, \varphi\big|'_{Y_{n_j}}(u_{n_j}) \to 0.$$

For n big enough, we have

$$
\begin{aligned}
c + 1 + ||u_{n_j}|| &\ge \varphi(u_{n_j}) - \frac{1}{p}\langle\varphi'(u_{n_j}), u_{n_j}\rangle \\
&= (\frac{1}{2} - \frac{1}{p})||u_{n_j}||^2 + (\frac{\mu}{p} - \frac{\mu}{q})|u_{n_j}|_q^q \\
&\ge (\frac{1}{2} - \frac{1}{p})||u_{n_j}||^2 + (\frac{\mu}{p} - \frac{\mu}{q})\beta_0^q||u_{n_j}||^q.
\end{aligned}
$$

Thus (u_{n_j}) is bounded in H_0^1. Going if necessary to a subsequence, we can assume that $u_{n_j} \rightharpoonup u$ in H_0^1. It is easy to conclude, as in Lemma 2.17, that $u_{n_j} \to u$ in H_0^1 and $\varphi'(u) = 0$. $\square$

3.7   Concave and critical nonlinearities

This section is devoted to the problem

$$(\mathcal{P}_4) \qquad \begin{cases} -\Delta u = |u|^{2^*-2}u + \mu|u|^{q-2}u, \\ u \in H_0^1(\Omega), \end{cases}$$

where Ω is a bounded domain of $\mathbb{R}^N$, $N \ge 3$, and $1 < q < 2$. The energy is defined on $H_0^1(\Omega)$ by

$$\varphi_\mu(u) := \int_\Omega \Big[\frac{|\nabla u|^2}{2} - \frac{|u|^{2^*}}{2^*} - \mu\frac{|u|^q}{q}\Big]dx.$$

We choose an orthonormal basis (e_j) of $H_0^1(\Omega)$ and we define $X_j := \mathbb{R}e_j$.

Lemma 3.21. *There exists $k > 0$ such that, for any $\mu > 0$ and*

$$\text{(3.6)} \qquad c < S^{N/2}/N - k\,\mu^{2^*/(2^*-q)},$$

the functional φ_μ satisfies the $(PS)_c^*$ condition.

Proof. 1) Consider a sequence $(u_{n_j}) \subset H_0^1$ such that

$$n_j \to \infty,\, u_{n_j} \in Y_{n_j},\, \varphi(u_{n_j}) \to c,\, \varphi'\big|_{Y_{n_j}}(u_{n_j}) \to 0,$$

where $\varphi := \varphi_\mu$ and c satisfies (3.6).

As in the proof of Theorem 3.20, (u_{n_j}) is bounded in H_0^1. Going if necessary to a subsequence, we can assume that

$$\begin{aligned}
u_{n_j} &\rightharpoonup u \quad \text{in } H_0^1, \\
u_{n_j} &\to u \quad \text{in } L^q, \\
u_{n_j} &\to u \quad \text{a.e. on } \Omega.
\end{aligned}$$

Define $f(u) := |u|^{2^*-2}u$. Since (u_{n_j}) is bounded in L^{2^*}, $(f(u_{n_j}))$ is bounded in $L^{2N/(N+2)}$ and so (see [90])

$$f(u_{n_j}) \rightharpoonup f(u) \quad \text{in} \quad L^{2N/(N+2)}.$$

It follows that

$$-\Delta u = |u|^{2^*-2}u + \mu|u|^{q-2}u$$

and

$$\text{(3.7)} \qquad \varphi(u) = \left(\frac{1}{2} - \frac{1}{2^*}\right)|u|_{2^*}^{2^*} + \left(\frac{1}{2} - \frac{1}{q}\right)\mu|u|_q^q.$$

2) We write $v_{n_j} := u_{n_j} - u$. Brézis-Lieb lemma leads to

$$|u_{n_j}|_{2^*}^{2^*} = |u|_{2^*}^{2^*} + |v_{n_j}|_{2^*}^{2^*} + o(1).$$

Hence we obtain

$$\text{(3.8)} \qquad \varphi(u) + \frac{|\nabla v_{n_j}|_2^2}{2} - \frac{|v_{n_j}|_{2^*}^{2^*}}{2^*} \to c.$$

Since $\langle \varphi'(u_{n_j}), u_{n_j} \rangle \to 0$, we obtain also

$$\begin{aligned}
|\nabla v_{n_j}|_2^2 - |v_{n_j}|_{2^*}^{2^*} &\to |u|_q^q + |u|_{2^*}^{2^*} - |\nabla u|_2^2 \\
&= -\langle \varphi'(u), u \rangle \\
&= 0.
\end{aligned}$$

We may therefore assume that

$$|\nabla v_{n_j}|_2^2 \to b, \quad |v_{n_j}|_{2^*}^{2^*} \to b.$$

By the Sobolev inequality, we have

$$|\nabla v_{n_j}|_2^2 \geq S|v_{n_j}|_{2^*}^2.$$

and so $b \geq Sb^{2/2^*}$. If $b = 0$, the proof is complete. Assume $b \geq S^{N/2}$. We obtain from (3.7) and (3.8)

$$
\begin{aligned}
c &= (\frac{1}{2} - \frac{1}{2^*})(b + |u|_{2^*}^{2^*}) + (\frac{1}{2} - \frac{1}{q})\mu|u|_q^q \\
&\geq \frac{1}{N}(S^{N/2} + |u|_{2^*}^{2^*}) + (\frac{1}{2} - \frac{1}{q})\mu|\Omega|^{(2^*-q)/2^*}|u|_{2^*}^q \\
&= S^{N/2}/N + |u|_{2^*}^{2^*}/N - a\mu|u|_{2^*}^q.
\end{aligned}
$$

If we define $k \geq 0$ by

$$\min_{t>0}(t^{2^*}/N - a\mu t^q) = -k\mu^{2^*/(2^*-q)},$$

we contradict (3.6). $\square$

Theorem 3.22. (Garcia-Peral, 1991). *Assume that Ω is bounded and $1 < q < 2$. There exists $\mu^* > 0$ such that, for every $0 < \mu < \mu^*$, problem $(\mathcal{P}_4)$ has a sequence of solutions (v_k) such that $\varphi_\mu(v_k) < 0$ and $\varphi_\mu(v_k) \to 0$, $k \to \infty$.*

Proof. By the preceding lemma, there exists $\mu^* > 0$ such that, for every $0 < \mu < \mu^*$ and $c < 0$, the functional φ_μ satisfies the $(PS)_c^*$ condition. It suffices then to verify the other assumptions of Theorem 3.18 as in the proof of Theorem 3.20. $\square$

Chapter 4

Nehari manifold

4.1 Definition of Nehari manifold

Assume that $\varphi \in C^1(X, \mathbb{R})$ is such that $\varphi'(0) = 0$. A necessary condition for $u \in X$ to be a critical point of φ is that $\langle \varphi'(u), u \rangle = 0$. This condition defines the *Nehari manifold*

$$N := \{u \in X : \langle \varphi'(u), u \rangle = 0, u \neq 0\}.$$

A critical point $u \neq 0$ of φ is a *ground state* or a *least energy critical point* if

$$\varphi(u) = \inf_N \varphi.$$

In the next section, following Rabinowitz [72], we give a minimax characterization of $\inf_N \varphi$. It is simpler to prove that the minimax value is a critical value. On the other hand, it is easier to obtain the qualitative properties of $\inf_N \varphi$. Finally, we use the Nehari manifold to prove the existence of nodal solutions.

4.2 Ground states

In this section, we consider the problem

(B_1)
$$\begin{cases} -\Delta u + u = f(|x|, u), \\ u \in H_0^1(\Omega), \end{cases}$$

where Ω is a rotationally symmetric domain of $\mathbb{R}^N$, $N \geq 2$. We fix $0 \leq \rho < \sigma \leq \infty$ and we define

$$\Omega = \Omega(\rho, \sigma) := \text{int}\{x \in \mathbb{R}^N : \rho \leq |x| < \sigma\},$$

$$F(r, u) := \int_0^u f(r, s)ds.$$

We assume the following hypotheses :

(f_1) $f \in \mathcal{C}([0, \infty[\times\mathbb{R})$ and, for some $2 < p < 2^*$, $c_0 > 0$

$$|f(r, u)| \leq c_0(|u| + |u|^{p-1}),$$

(f_2) there exists $\alpha > 2$ such that, for every r and $u \in \mathbb{R}$

$$\alpha F(r, u) \leq u f(r, u),$$

(f_3) $f(r, u) = o(|u|)$, $|u| \to 0$, uniformly on $\mathbb{R}^+$,
(f_4) there exists $R > 0$ such that

$$\inf_{\substack{r>0 \\ |u|>R}} F(r, u) > 0,$$

(f_5) $f(r, u)/|u|$ is an increasing function of u on $\mathbb{R}\backslash\{0\}$ for every $r > 0$.
The radial solutions of $(\mathcal{P}_1)$ are the critical points of the functional

$$\varphi(u) := \int_\Omega \left[\frac{|\nabla u|^2}{2} + \frac{u^2}{2} - F(|x|, u)\right] dx$$

defined on the space $X := H^1_{0, \mathbf{O}(N)}(\Omega)$.

Lemma 4.1. *Under the above assumptions, for any $u \in X\backslash\{0\}$ there exists a unique $t(u) > 0$ such that $t(u)u \in N$. The maximum of $\varphi(tu)$ for $t \geq 0$ is achieved at $t = t(u)$. The function*

$$X\backslash\{0\} \to]0, \infty[: u \mapsto t(u)$$

is continuous and the map $u \to t(u)u$ defines a homeomorphism of the unit sphere of X with N.

   Proof. After integrating, we obtain from (f_2) the existence of $C_0 > 0$ such that

$$C_0(|u|^\alpha - 1) \leq F(r, u).$$

   Let $u \in X\backslash\{0\}$ be fixed and define the function $g(t) := \varphi(tu)$ on $[0, \infty[$. Clearly we have

(4.1) $g'(t) = 0 \iff u \in N$

$$\iff ||u||_1^2 = \frac{1}{t}\int f(x, tu)u\, dx.$$

By (f_5), the right hand side is an increasing function of t. It is easy to verify, using (f_{2-3-4}) that $g(0) = 0$, $g(t) > 0$ for $t > 0$ small and $g(t) < 0$

for t large. Therefore $\max\limits_{[0,\infty[} g$ is achieved at a unique $t = t(u)$ so that $g'(t(u)) = 0$ and $t(u)u \in N$. To prove the continuity of $t(u)$, assume that $u_n \to u$ in $X\backslash\{0\}$. It is easy to verify that $(t(u_n))$ is bounded. If a subsequence of $(t(u_n))$ converges to t_0, it follows from (4.1) that $t_0 = t(u)$. But then $t(u_n) \to t(u)$. Finally the continuous map from the unit sphere of X to N, $u \mapsto t(u)u$, is inverse to the retraction $u \mapsto u/\|u\|$. $\square$

We define

$$
\begin{aligned}
c_1 &:= \inf_N \varphi, \\
c_2 &:= \inf_{\substack{u \in X \\ u \neq 0}} \max_{t \geq 0} \varphi(tu), \\
c &:= \inf_{\gamma \in \Gamma} \max_{t \in [0,1]} \varphi(\gamma(t)),
\end{aligned}
$$

where

$$\Gamma := \{\gamma \in \mathcal{C}([0,1], X) : \gamma(0) = 0, \varphi(\gamma(1)) < 0\}.$$

Theorem 4.2. *Under the assumptions (f_1)-(f_5), we have that*

$$c_1 = c_2 = c > 0$$

and c is a critical value of φ.

Proof. 1) The preceding lemma implies that $c_1 = c_2$. Since $\varphi(tu) < 0$ for $u \in X\backslash\{0\}$ and t large, we obtain $c \leq c_2$. The manifold N separates X into two components. By (f_1) and (f_3) the component containing the origin also contains a small ball around the origin. Moreover $\varphi(u) \geq 0$ for all u in this component, because $\langle \varphi'(tu), u \rangle \geq 0$ for all $0 \leq t \leq t(u)$. Thus every $\gamma \in \Gamma$ has to cross N and $c_1 \leq c$.

2) In order to prove that c is a critical value of φ, we apply Theorem 2.9 with $M := [0,1]$, $M_0 := \{0,1\}$ and

$$\Gamma_0 := \{\gamma_0 : \{0,1\} \to X : \gamma_0(0) = 0, \varphi(\gamma_0(1)) < 0\}.$$

By (f_1) and (f_3) there exists $r > 0$ such that

$$\min_{\|u\| \leq r} \varphi(u) = 0, \qquad \inf_{\|u\|=r} \varphi(u) > 0.$$

Hence we obtain

$$c \geq \inf_{\|u\|=r} \varphi(u) > 0 = \sup_{\gamma_0 \in \Gamma_0} \sup_{u \in M_0} \varphi(\gamma_0(u)).$$

By a variant of Lemma 3.11, φ satisfies $(PS)_c$ condition and the proof is complete. $\square$

The following result is independent of the Palais-Smale condition.

Theorem 4.3. *Under the assumptions (f_1)-(f_5), if $v \in N$ and $\varphi(v) = c$ then v is a critical point of φ.*

Proof. Assume that $v \in N$, $\varphi(v) = c$ and $\varphi'(v) \neq 0$. Then there exists $\delta > 0$, $\lambda > 0$ such that

$$|u - v| \leq 3\delta \Rightarrow \|\varphi'(u)\| \geq \lambda.$$

For $\varepsilon := \min\{c/2, \lambda\delta/8\}$, $S := B(v, \delta)$ Lemma 2.3 yields a deformation η such that
a) $\eta(1, u) = u$ if $u \notin \varphi^{-1}([c - 2\varepsilon, c + 2\varepsilon])$,
b) $\eta(1, \varphi^{c+\varepsilon} \cap B(v, \delta)) \subset \varphi^{c-\varepsilon}$,
c) $\varphi(\eta(1, u)) \leq \varphi(u)$, $\forall u \in X$.
It is clear that $\max_{t>0} \varphi(\eta(1, tv)) < c_2 = c$ which contradicts the definition of c. $\square$

4.3　Properties of critical values

In this section we denote by $c(\rho, \sigma)$ the critical value $\inf_N \varphi$ corresponding to $\Omega(\rho, \sigma)$.

Proposition 4.4. *Under the assumptions (f_1)-(f_5), $c(\rho, \sigma)$ has the following properties :*
a) *if $0 \leq \rho \leq \rho' < \sigma' \leq \sigma \leq \infty$, then $c(\rho, \sigma) \leq c(\rho', \sigma')$,*
b) *$c(\rho, \sigma) \to \infty$ as $\sigma - \rho \to 0$,*
c) *$c(\rho, \infty) \to \infty$ as $\rho \to \infty$,*
d) *$c(\rho, \sigma)$ is lower semi-continuous with respect to (ρ, σ).*

Proof. a) If $0 \leq \rho \leq \rho' < \sigma \leq \sigma < \infty$, then $N(\rho', \sigma')$ is contained in $N(\rho, \sigma)$, so that $c(\rho, \sigma) \leq c(\rho', \sigma')$.
b) By (f_2), every $u \in N(\rho, \sigma)$ satisfies

$$(4.2) \qquad \varphi(u) \geq \frac{1}{2}\|u\|_1^2 - \frac{1}{\alpha} \int_{\Omega(\rho,\sigma)} f(|x|, u)u \, dx \geq \left(\frac{1}{2} - \frac{1}{\alpha}\right)\|u\|_1^2.$$

Using (f_1) and (f_3), we obtain, for $u \in N(\rho, \sigma)$,

$$\|u\|_1^2 = \int_{\Omega(\rho,\sigma)} f(|x|, u)u \, dx \leq \frac{1}{2}|u|_2^2 + c_0|u|_p^p$$

$$\leq \frac{1}{2}|u|_2^2 + a_0|u|_{2*}^p(\sigma - \rho)^{1-p/2^*}$$

$$\leq \frac{1}{2}\|u\|_1^2 + a_1\|u\|_1^p(\sigma - \rho)^{1-p/2^*}$$

so that

$$a_2(\sigma - \rho)^{-1+p/2^*} \le \|u\|_1^{p-2}.$$

If $\sigma - \rho \to 0$, then $\|u\|_1 \to \infty$ and, by (4.2), $\varphi(u) \to \infty$, for $u \in N(\rho, \sigma)$.

c) By (f_1), (f_3) and the next lemma, every $u \in N(\rho, \infty)$ satisfies

$$\|u\|_1^2 = \int_{\Omega(\rho,\infty)} f(|x|, u)u\, dx \;\le\; \frac{1}{2}|u|_2^2 + c_0|u|_p^p$$

$$\le\; \frac{1}{2}|u|_2^2 + a_3|u|_2^2\|u\|_1^{p-2}\rho^{(1-N)(p-2)/2}$$

$$\le\; \frac{1}{2}\|u\|_1^2 + a_3\|u\|_1^p\rho^{(1-N)(p-2)/2}$$

so that

$$a_4\rho^{(N-1)/2} \le \|u\|_1.$$

If $\rho \to \infty$, then $\|u\|_1 \to \infty$ and by (4.2), $\varphi(u) \to \infty$, for $u \in N(\rho, \infty)$.

d) We have to show that

$$c(\rho, \sigma) \le \varliminf_{n \to \infty} c(\rho_n, \sigma_n) \text{ if } (\rho_n, \sigma_n) \to (\rho, \sigma), n \to \infty.$$

We only treat the case $\sigma_n \equiv \infty$, since the other cases are similar. We assume that $\rho_n \to \rho > 0$. For every n, there exists $u_n \in N(\rho_n, \infty)$ such that $\varphi(u_n) = c(\rho_n, \infty)$. We define v_n on $\Omega(\rho, \infty)$ by

$$v_n(x) := u_n(\rho_n x/\rho).$$

According to Lemma 4.1, there exists a unique $t_n > 0$ with $w_n := t_n v_n \in N(\rho, \infty)$. It is clear that $\varphi(w_n) \ge c(\rho, \infty)$. It remains to prove that

$$(4.3) \qquad\qquad \varphi(w_n) = \varphi(u_n) + o(1), \quad n \to \infty.$$

By definition of $N(\rho, \infty)$, we have

$$0 = \int_{\Omega(\rho,\infty)} \left[\frac{t_n^2}{2}(|\nabla v_n|^2 + v_n^2) - t_n v_n f(|x|, t_n v_n) \right] dx$$

$$= \int_{\Omega(\rho_n,\infty)} \left[\frac{t_n^2 \rho_n^2}{2\rho^2}|\nabla u_n|^2 + \frac{t_n^2}{2}u_n^2 - t_n u_n f(|x|, t_n u_n) \right] \frac{\rho^N}{\rho_n^N} dx$$

and, by definition of $N(\rho_n, \infty)$,

$$0 = \int_{\Omega(\rho_n,\infty)} \left[\frac{1}{2}(|\nabla u_n|^2 + |u_n^2|) - u_n f(|x|, u_n) \right] dx$$

and so $t_n = 1 + o(1)$. It follows that

$$\|u_n - w_n\| = o(1), \quad n \to \infty,$$

which implies (4.3). $\square$

Lemma 4.5. (Strauss inequality). *If $N \geq 2$, there exists $c(N) > 0$ such that, for every $u \in H^1_{\mathbf{O}(N)}(\mathbb{R}^N)$,*

$$|u(x)| \leq c(N)|u|_2^{1/2}|\nabla u|_2^{1/2}|x|^{(1-N)/2} \text{ a.e. on } \mathbb{R}^N.$$

Proof. By density, it suffices to consider $u \in H^1_{\mathbf{O}(N)}(\mathbb{R}^N) \cap \mathcal{D}(\mathbb{R}^N)$. Since

$$2ur^{N-1}\frac{d}{dr}u \leq \frac{d}{dr}(r^{N-1}u^2),$$

we obtain

$$\begin{aligned} r^{N-1}u^2(r) &\leq 2\int_r^\infty |u|\left|\frac{du}{dr}\right|s^{N-1}ds \\ &\leq c(N)^2|u|_2|\nabla u|_2. \quad \square \end{aligned}$$

4.4   Nodal solutions

In this section, we consider the problem

$(\mathcal{P}_2)$
$$\begin{cases} -\Delta u + u = f(|x|, u), \\ u \in H^1_0(\mathbb{R}^N). \end{cases}$$

A *node* of a radial solution of $(\mathcal{P}_2)$ is a radius $\rho > 0$ such that $u(x) = 0$ when $|x| = \rho$.

Theorem 4.6. (Bartsch-Willem, 1993). *Under assumptions (f_1)-(f_5), for every integer $k > 0$, there exists a pair u_k^+ and u_k^- of radial solutions of $(\mathcal{P}_2)$ with $u_k^-(0) < 0 < u_k^+(0)$, having exactly k nodes $0 < \rho_1^\pm < \ldots < \rho_k^\pm < \infty$.*

Proof. 1) We fix some integer $k \geq 1$ and want to find a radial solution u_k^+ of $(\mathcal{P}_1)$ having k nodes with $u_k^+(0) > 0$. We define, on $[0, \infty[\times \mathbb{R}$,

$$\begin{aligned} f^+(r, u) &:= f(r, u), & \text{if } u \geq 0, \\ &:= -f(r, -u), & \text{if } u < 0. \end{aligned}$$

The function φ_+ is defined on $X_{\rho,\sigma} := H^1_{0,\mathbf{O}(N)}(\Omega_{\rho,\sigma})$ by

$$\varphi_+(u) := \int_{\Omega_{\rho,\sigma}} \left[\frac{|\nabla u|^2}{2} + \frac{u^2}{2} - F^+(|x|, u)\right] dx$$

where $F^+(r, .)$ is the primitive of $f^+(r, .)$ satisfying $F^+(r, 0) = 0$. By Theorem 4.2,

$$c^+(\rho, \sigma) := \inf_{N_{\rho,\sigma}^+} \varphi_+,$$

where

$$N^+_{\rho,\sigma} := \{u \in X_{\rho,\sigma} : \langle \varphi'_+(u), u \rangle = 0, u \neq 0\},$$

is a critical value of φ_+. Let u be the corresponding critical point. Since $\varphi_+(|u|) = \varphi_+(u)$ and

$$\langle \varphi'_+(|u|), |u| \rangle = \langle \varphi'_+(u), u \rangle = 0,$$

it follows from Theorem 4.3 that $|u|$ is also a critical point of φ_+. Thus we can assume that $|u| = u$. In particular u is a nonnegative critical point of φ.

The non-positive critical point of φ is obtained in a similar way by considering

$$\begin{aligned} f_-(r, u) &:= -f(r, -u) \quad \text{if} \quad u \geq 0, \\ &:= f(r, u) \quad \text{if} \quad u < 0. \end{aligned}$$

The corresponding critical value is $c^-(\rho, \sigma)$.

2) We define

$$C^+(\rho_1, \ldots, \rho_k) := \sum_{j=0}^{k} c^{\varepsilon_j}(\rho_j, \rho_{j+1}),$$

for

$$0 = \rho_0 < \rho_1 < \ldots < \rho_k < \rho_{k+1} = \infty,$$

where

$$\begin{aligned} \varepsilon_j &= +, \quad j \text{ even}, \\ &= -, \quad j \text{ odd}. \end{aligned}$$

By Proposition 4.4, C^+ attains its infimum at some point $(\rho_1, \ldots, \rho_k)$. For $0 \leq j \leq k$ and j even, there exists a nonnegative solution u_j of

$$(4.4) \qquad \begin{cases} -\Delta u + u = f(|x|, u), \\ u \in X_{\rho_j, \rho_{j+1}}, \end{cases}$$

such that $\varphi(u_j) = c^+(\rho_j, \rho_{j+1})$. And for j odd, there exists a nonpositive solution u_j of (4) such that $\varphi(u_j) = c^-(\rho_j, \rho_{j+1})$. We define on $\mathbb{R}^N$

$$u_k^+(x) := u_j(x), \text{ if } \rho_j \leq |x| < \rho_{j+1}.$$

3) By Strauss inequality, u_k^+ is continuous except perhaps at 0. Brézis-Kato theorem implies that $u_k^+ \in L^p_{\text{loc}}(\Omega_{0,\rho_1})$ for all $1 \leq p < \infty$. Thus $u \in W^{2,p}_{\text{loc}}(\Omega_{0,\rho_1})$ for all $1 \leq p < \infty$. In particular $u \in C^2(\Omega_{0,\rho_1})$ and, by the maximum principle, $u_k^+ > 0$ on $\Omega_{\rho_j, \rho_{j+1}}$, j even, and $u_k^+ < 0$ on $\Omega_{\rho_j, \rho_{j+1}}$, j odd. It follows from the next lemma that u_k^+ is a solution of $(\mathcal{P}_2)$. $\square$

Lemma 4.7. *The function u_k^+ is a solution of*

(4.5) $$-\Delta u + u = f(|x|, u).$$

Proof. We define $u := u_k^+$. Clearly u satisfies (4.5) in $\{x \in \mathbb{R}^N : |x| \neq \rho_j, j = 1, \dots, k\}$. Since x is away from the origin we set $r := |x|$ and treat (4.5) as an ordinary differential equation. Thus we write $u(r)$ instead of $u(x)$. We know already that u is of class $\mathcal{C}^2$ on

$$U := \{r > 0 : r \neq \rho_j \text{ for } j = 1, \dots, k\}$$

and satisfies

(4.6) $$-(r^{N-1}u')' = r^{N-1}(f(r, u) - u)$$

on U. Here $'$ denotes d/dr, of course. We have to show that u satisfies (4.6) on all of $\mathbb{R}^+$. This is the case if and only if

$$u'_+ := \lim_{r \searrow \rho_j} u'(r) = \lim_{r \nearrow \rho_j} u'(r) =: u'_-.$$

We prove this by contradiction. Assume $u'_+ \neq u'_-$ and set $\rho := \rho_{j-1}, \sigma := \rho_j, \tau := \rho_{j+1}$. We may assume that $u \geq 0$ on $[\rho, \sigma]$ and $u \leq 0$ on $[\sigma, \tau]$. Now we fix $\delta > 0$ and define $v : [\rho, \tau] \to \mathbb{R}$ by

$$v(r) := u(r) \text{ if } |r - \sigma| \geq \delta,$$

$$v(r) := u(\sigma - \delta) + (r - \sigma + \delta)(u(\sigma + \delta) - u(\sigma - \delta))/2\delta$$

if $|r - \sigma| \leq \delta$. Clearly, v is continuous on $[\rho, \tau]$. Let $\sigma_0 = \sigma_0(\delta) \in (\sigma - \delta, \sigma + \delta)$ be defined by $v(\sigma_0) = 0$. According to Lemma 4.1, there exist $\alpha = \alpha(\delta) > 0$ and $\beta = \beta(\delta) > 0$ such that

$$\alpha v \in N^+(\rho, \sigma_0), \quad \beta v \in N^-(\sigma_0, \tau).$$

Next we define $w : [\rho, \tau] \to \mathbb{R}$ by setting

$$w(r) := \begin{cases} \alpha v(r) & \text{if } \rho \leq r \leq \sigma_0, \\ \beta v(r) & \text{if } \sigma_0 \leq r \leq \tau. \end{cases}$$

By construction of u we obtain $\psi(u) \leq \psi(w)$ where

$$\psi(h) := \int_\rho^\tau \left(\frac{1}{2}h'^2 + \frac{1}{2}h^2 - F(r, h) \right) r^{N-1} dr.$$

Now, we have to use the convexity assumption (f_5) again. It implies that $F(r, \sqrt{h})$ and $F(r, -\sqrt{h})$ are convex function of $h \in \mathbb{R}^+$ because $\frac{\partial}{\partial h} F(r, \pm\sqrt{h})$ is increasing. Hence we obtain

$$F(r, w) \geq F(r, u) + \frac{w^2 - u^2}{2} \cdot \frac{f(r, u)}{u} \text{ if } u, w > 0.$$

It follows that

$$\left[\int_\rho^{\sigma-\delta} + \int_{\sigma+\delta}^\tau\right]\left(\frac{1}{2}w'^2 + \frac{1}{2}w^2 - F(r,w)\right)r^{N-1}dr$$

$$\leq \left[\int_\rho^{\sigma-\delta} + \int_{\sigma+\delta}^\tau\right]\left(\frac{1}{2}w'^2 + \frac{1}{2}w^2 - F(r,u) - \frac{w^2 - u^2}{2}\cdot\frac{f(r,u)}{u}\right)r^{N-1}dr.$$

On the other hand, we have $u \cdot f(r,u) \geq 0$ by (f_3) and (f_5); hence

$$\int_{\sigma-\delta}^{\sigma+\delta}\left(\frac{1}{2}w'^2 + \frac{1}{2}w^2 - F(r,w)\right)r^{N-1}dr$$

$$\leq \int_{\sigma-\delta}^{\sigma+\delta}\left(\frac{1}{2}w'^2 + \frac{1}{2}w^2 - F(r,u) + \frac{1}{2}uf(r,u) - F(r,w) + F(r,u)\right)r^{N-1}dr.$$

Since, by the definition of u,

$$\int_\rho^\tau (u'^2 + u^2)r^{N-1}dr = \int_\rho^\tau f(r,u)ur^{N-1}dr,$$

we obtain

(4.7) $\psi(w) \leq \psi(u) + \left[\int_\rho^{\sigma-\delta} + \int_{\sigma+\delta}^\tau\right]\left(\frac{1}{2}w'^2 + \frac{1}{2}w^2 - \frac{w^2}{2u}f(r,u)\right)r^{N-1}dr$

$$+ \int_{\sigma-\delta}^{\sigma+\delta}\left(\frac{1}{2}w'^2 + \frac{1}{2}w^2 - F(r,w) + F(r,u)\right)r^{N-1}dr.$$

Using (4.6) we see that

(4.8) $$\int_\rho^{\sigma-\delta}\left(\frac{1}{2}w'^2 + \frac{1}{2}w^2 - \frac{w^2}{2u}f(r,u)\right)r^{N-1}dr$$

$$= \frac{\alpha^2}{2}\int_\rho^{\sigma-\delta}(u'^2 + u^2 - uf(r,u))r^{N-1}dr = \frac{\alpha^2}{2}(\sigma-\delta)^{N-1}u'(\sigma-\delta)u(\sigma-\delta).$$

Moreover, since $u(\sigma) = 0$ and $(r^{N-1}u')(\sigma) = 0$ by (4.6), we obtain

$$u(\sigma - \delta) = -\delta u'_- + o(\delta),$$

(4.9)

$$(r^{N-1}u')'(\sigma - \delta) = \delta^{N-1}u'_- + o(\delta).$$

It is not difficult to verify that

(4.10) $$\lim_{\delta\searrow 0}\alpha(\delta) = 1 = \lim_{\delta\searrow 0}\beta(\delta).$$

It follows from (4.8), (4.9) and (4.10) that

$$(4.11) \quad \int_\rho^{\sigma-\delta} \left(\frac{1}{2}w'^2 + \frac{1}{2}w^2 - \frac{w^2}{2u} f(r,u) \right) r^{N-1} dr = -\frac{\sigma^{N-1}}{2} u_-'^2 \delta + o(\delta).$$

Similarly we can prove that

$$(4.12) \quad \int_{\sigma+\delta}^\rho \left(\frac{1}{2}w'^2 + \frac{1}{2}w^2 - \frac{w^2}{2u} f(r,u) \right) r^{N-1} dr = -\frac{\sigma^{N-1}}{2} u_+'^2 \delta + o(\delta).$$

Next one checks that

$$(4.13) \quad \int_{\sigma-\delta}^{\sigma+\delta} \left(\frac{1}{2}w^2 - F(r,w) + F(r,u) \right) r^{N-1} dr = o(\delta),$$

and that

$$(4.14) \quad \int_{\sigma-\delta}^{\sigma+\delta} \frac{1}{2}w'^2 r^{N-1} dr = \frac{\delta^{N-1}}{4}(u_+' + u_-')^2 \delta + o(\delta).$$

The last equality comes from (4.10) and

$$
\begin{aligned}
\int_{\sigma-\delta}^{\sigma+\delta} \frac{1}{2} v'^2 r^{N-1} dr &= \frac{(u(\sigma+\delta) - u(\sigma-\delta))^2}{8\delta^2} \cdot \left(\frac{(\delta+\delta)^N}{N} \cdot \frac{(\sigma-\delta)^N}{N} \right) \\
&= \frac{\sigma^{N-1}}{4}(u_+' + u_-')^2 \delta + o(\delta).
\end{aligned}
$$

Now we obtain a contradiction, as follows. Using (4.7) and (4.11)-(4.14) we have

$$\psi(w) \leq \psi(u) - \frac{\sigma^{N-1}}{4}(u_+' - u_-')^2 \delta + o(\delta).$$

This implies that $\psi(w) < \psi(u)$ for $\delta > 0$ small enough, which contradicts $\psi(u) \leq \psi(w)$. $\square$

Corollary 4.8. *Assume that $N \geq 2, p, q \in]2, 2^*[$ and $\lambda, \mu \in]0, \infty[$. Then, for every integer $k \geq 0$, there exists a pair u_k^+ and u_k^- of radial solutions of*

$$
\begin{cases}
-\Delta u + u = \lambda(u^+)^{p-1} - \mu(u^-)^{q-1}, \\
u \in H^1(\mathbb{R}^N),
\end{cases}
$$

with $u_k^-(0) < 0 < u_k^+(0)$, having exactly k nodes.

Chapter 5

Relative category

5.1   Category

We identify the two dimensional torus $\mathbb{T}^2$ with the quotient space $\mathbb{R}^2/\mathbb{Z}^2$. We consider a function $\varphi \in \mathcal{C}^1(\mathbb{T}^2, \mathbb{R})$ having a maximum at M and a minimum at m. We also assume that the level sets of φ are the curves on the figure below

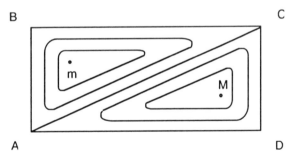

The function φ has only three critical points on the torus: M, m and the point corresponding to A, B, C and D. On the next figure, we give a covering of the torus by three closed contractible sets.

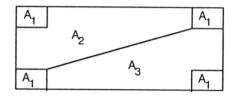

Following Lusternik and Schnirelman, we define the *category* $\mathrm{cat}_X(A)$ of a closed subset A of a topological space X as the least integer n such that there exists a covering of A by n closed sets contractible in X. The number of critical points of a C^1 functional φ defined on a compact manifold X is greater or equal to $\mathrm{cat}_X(X)$. The corresponding critical values are given by

$$c_k := \inf_{A \in \mathcal{A}_k} \sup_{u \in A} \varphi(u),$$

$$\mathcal{A}_k := \{A \subset X : A \text{ is closed, } \mathrm{cat}_X(A) \geq k\}.$$

5.2 Relative category

We consider now the usual representation of the torus in $\mathbb{R}^3$ and the altitude φ.

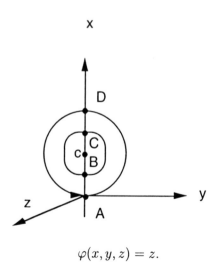

$$\varphi(x, y, z) = z.$$

The function φ has four critical points A, B, C and D. The set φ^c contains two critical points A and B and is invariant with respect to the negative gradient flow

$$\dot{\sigma} = -\nabla\varphi(\sigma).$$

The category $\mathrm{cat}_{\varphi^c}(\varphi^c)$ is equal to two.

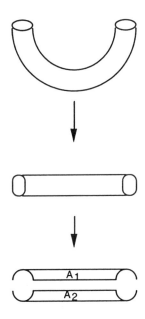

The set $X := \{u \in \mathbb{T}^2 : \varphi(u) \geq c\}$ contains two critical points C and D. On X, the negative gradient flow converges to C, D or $Y := \{u \in X : \varphi(u) = c\}$. There is a covering (A_0, A_1, A_2) of X such that A_1 and A_2 are contractible in X and there exists a continuous deformation of A_0 into Y.

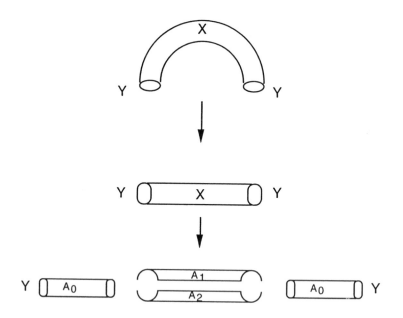

Let us now give some formal definitions.

Definition 5.1. *A closed subset A is contractible in a topological space X if there exists $h \in \mathcal{C}([0,1] \times A, X)$ such that, for every $u, v \in A$,*

$$h(0,u) = u, \quad h(1,u) = h(1,v).$$

Definition 5.2. *Let A, B, Y be closed subsets of a topological space X. Then, by definition, $A \prec_Y B$ in X if $Y \subset A \cap B$ and there exists $h \in \mathcal{C}([0,1] \times A, X)$ such that*
a) $h(0,u) = u$, $h(1,u) \in B$, $\forall u \in A$,
b) $h(t, Y) \subset Y$, $\forall t \in [0,1]$.

Definition 5.3. *Let $Y \subset A$ be closed subsets of a topological space X. The category of A in X relative to Y is the least integer n such that there exists $n + 1$ closed subsets $A_0, A_1, \ldots, A_n$ of X satisfying*
a) $A = \bigcup\limits_{j=0}^{n} A_j,$
b) $A_1, \ldots, A_n$ are contractible in X,
c) $A_0 \prec_Y Y$ in X.
We denote the category of A in X relative to Y by $\mathrm{cat}_{X,Y}(A)$.

Definition 5.4. *Let A be a closed subset of a topological space X. The category of A in X is defined by* $\mathrm{cat}_X(A) := \mathrm{cat}_{X,\phi}(A)$.

Let us now prove some elementary properties of relative category.

Lemma 5.5. *Let A, B, C, Y be closed subsets of X such that $Y \subset A \cap B \cap C$. If $A \prec_Y B$ and $B \prec_Y C$ in X then $A \prec_Y C$ in X.*

Proof. Assume that $A \prec_Y B$ and $B \prec_Y C$ by means of deformations h and g. Using the deformation

$$
\begin{aligned}
(g * h)(t, u) &= h(2t, u), \quad 0 \le t \le 1/2, u \in A, \\
&= g(2t - 1, h(1, u)), \quad 1/2 < t \le 1, u \in A,
\end{aligned}
$$

we prove that $A \prec_Y C$ in X. $\square$

Proposition 5.6. *Let A, B, Y be closed subsets of X such that $Y \subset A$. The relative category satisfies the following properties:*
a) normalisation: $\mathrm{cat}_{X,Y}(Y) = 0$,
b) subadditivity: $\mathrm{cat}_{X,Y}(A \cup B) = \mathrm{cat}_{X,Y}(A) + \mathrm{cat}_X(B)$,
c) monotonicity: if $A \prec_Y B$ then $\mathrm{cat}_{X,Y}(A) \le \mathrm{cat}_{X,Y}(B)$.

Proof. Properties a) and b) are obvious. Let us prove c). Assume that $A \prec_Y B$ by means of deformation h and let $(B_0, \ldots, B_n)$ be the covering of B corresponding to $n := \mathrm{cat}_{X,Y}(B)$. Define

$$
A_j := \{u \in A : h(1, u) \in B_j\}, \quad j = 0, \ldots, n.
$$

It is clear that $A = \bigcup_{j=0}^n A_j$, $A_0 \prec_Y B_0$, $A_j \prec_\phi B_j$, $j = 1, \ldots, n$. The preceding lemma implies that $\mathrm{cat}_{X,Y}(A) \le n$. $\square$

The proof of the following result is left to the reader.

Proposition 5.7. *Let $\Phi_{X,Y}$ be a function defined on the class of closed subsets of X containing Y. If $\Phi_{X,Y}$ satisfies properties a), b), c) of the preceding proposition, then $\Phi_{X,Y} \le \mathrm{cat}_{X,Y}$.*

Definition 5.8. *A metric space X is an absolute neighborhood extensor, shortly an ANE, if for every metric space E, every closed subset F of E and every continuous map $f : F \to X$, there exists a continuous extension of f defined on a neighborhood of F in E.*

Proposition 5.9. *Let A be a closed subset of an ANE X. There exists a closed neighborhood B of A in X such that $\mathrm{cat}_X(B) = \mathrm{cat}_X(A)$.*

Proof. It suffices to consider the case where $\text{cat}_X(A) = 1$. Let h be the corresponding deformation. The set

$$N := ([0,1] \times A) \cup (\{0,1\} \times X)$$

is closed in $M := [0,1] \times X$. The map $f : N \to X$ defined by

$$\begin{aligned}
f(t,u) &:= h(t,u), & t &\in [0,1], u \in A, \\
&:= u, & t &= 0, u \in X, \\
&:= h(1,u_0), & t &= 1, u \in X,
\end{aligned}$$

when $u_0 \in X$ is fixed, is continuous. By assumption, there exists a continuous extension g of f defined on a neighborhood U of N. Since X is normal, we can assume that U is closed. The compactness of $[0,1]$ implies the existence of a closed neighborhood B of A such that $[0,1] \times B \subset U$. But then B is contractible in X and $\text{cat}_X(B) = 1$. $\square$

5.3   Quantitative deformation lemma

In this section we consider, for simplicity, the following situation.

(A) X is a Banach space, $\psi \in C^2(X,\mathbb{R})$, $V := \{v \in X : \psi(v) = 1\}$, for every $v \in V$, $\psi'(v) \neq 0$.

Definition 5.10. *The tangent space of V at v is defined by*

$$T_v V := \{y \in X : \langle \psi'(v), y \rangle = 0\}.$$

Let $\varphi \in C^1(X,\mathbb{R})$ and $v \in V$. The norm of the derivative of the restriction of φ to V at v is defined by

$$\|\varphi'(v)\|_* := \sup_{\substack{y \in T_v V \\ \|y\|=1}} \langle \varphi'(v), y \rangle.$$

The point v is a critical point of the restriction of φ to V if the restriction of $\varphi'(v)$ to $T_v V$ is equal to 0. We define also

$$\varphi^d := \{v \in V : \varphi(v) \leq d\}$$

and

$$K_c := \{u \in V : \varphi(u) = c, \|\varphi'(u)\|_* = 0\}.$$

We will use the following *duality lemma.*

Lemma 5.11. *If $f, g \in X'$, then*

$$\sup_{\substack{\langle g,y \rangle = 0 \\ \|y\|=1}} \langle f, y \rangle = \min_{\lambda \in \mathbb{R}} \|f - \lambda g\|.$$

Proof. a) It is clear that, for every $\lambda \in \mathbb{R}$,

$$\begin{aligned} \sup_{\substack{\langle g,y \rangle = 0 \\ \|y\|=1}} \langle f, y \rangle &\leq \sup_{\|y\|=1} \langle f - \lambda g, y \rangle \\ &= \|f - \lambda g\|. \end{aligned}$$

b) By the Hahn-Banach theorem, there is a continuous linear functional $\tilde{f}$ on X agreeing with f on ker g and such that

$$\sup_{\substack{\|y\|=1 \\ \langle g,y \rangle = 0}} \langle f, y \rangle = \|\tilde{f}\|.$$

Since $\ker(f - \tilde{f}) \subset \ker g$, there is $\lambda \in \mathbb{R}$ such that $f - \tilde{f} = \lambda g$. Hence we obtain

$$\sup_{\substack{\|y\|=1 \\ \langle g,y \rangle = 0}} \langle f, y \rangle = \|\tilde{f}\| = \|f - \lambda g\|. \qquad \square$$

Proposition 5.12. *If $\varphi \in C^1(X, \mathbb{R})$ and $u \in V$ then*

$$\|\varphi'(u)\|_* = \min_{\lambda \in \mathbb{R}} \|\varphi'(u) - \lambda \psi'(u)\|.$$

In particular, u is a critical point of $\varphi\big|_V$ if and only if there exists $\lambda \in \mathbb{R}$ such that

$$\varphi'(u) = \lambda \psi'(u).$$

Definition 5.13. *Let $\varphi \in C^1(X, \mathbb{R})$. A tangent pseudo-gradient vector field for φ on*

$$M := \{u \in V : \|\varphi'(u)\|_* \neq 0\}$$

is a locally Lipschitz continuous vector field $g : M \to X$ such that, for every $u \in M$, $g(u) \in T_u V$ and

$$\|g(u)\| \leq 2\|\varphi'(u)\|_*,$$

$$\langle \varphi'(u), g(u) \rangle \geq \|\varphi'(u)\|_*^2.$$

Lemma 5.14. *Let $\varphi \in C^1(X, \mathbb{R})$. There exists a tangent pseudo-gradient vector field for φ on M.*

Proof. For every $v \in M$, there exists $x \in T_v V$ such that $||x|| = 1$ and

$$\langle \varphi'(v), x \rangle > \frac{2}{3} ||\varphi'(v)||_*.$$

There exists also $z \in X$ such that $\langle \psi'(v), z \rangle = 1$. Define $y := \frac{3}{2}||\varphi'(v)||_* x$ and, for $u \in V$ such that $\langle \psi'(u), z \rangle \neq 0$,

$$g_v(u) := y - \frac{\langle \psi'(u), y \rangle}{\langle \psi'(u), z \rangle} z.$$

Since $g_v(v) = y$, we obtain

$$||g_v(v)|| < 2||\varphi'(v)||_*, \quad \langle \varphi'(v), g_v(v) \rangle > ||\varphi'(v)||_*^2.$$

Since φ' and g_v are continuous, there exists an open neighborhood N_v of v such that, for every $u \in N_v$,

$$||g_v(u)|| \leq 2||\varphi'(u)||_*, \quad \langle \varphi'(u), g_v(u) \rangle \geq ||\varphi'(u)||_*^2.$$

The family $\mathcal{N} := \{N_v : v \in M\}$ is an open covering of M. Since M is metric, hence paracompact, there exists a locally finite open covering $\mathcal{M} = \{M_i : i \in I\}$ of M such that, for every $i \in I$, there exists $v \in V$ satisfying $\overline{M_i} \subset N_v$. We define

$$\begin{aligned} g_i(u) &:= g_v(u), \quad u \in N_v, \\ &:= 0, \quad u \notin N_v, \end{aligned}$$

and

$$\begin{aligned} \rho_i(u) &:= \operatorname{dist}(u, X \backslash M_i), \\ g(u) &:= \sum_{i \in I} \frac{\rho_i(u) g_i(u)}{\sum_{j \in I} \rho_j(u)}. \end{aligned}$$

It is easy to verify that g is a tangent pseudo-gradient vector field for φ on M. □

Lemma 5.15. *Let $\varphi \in \mathcal{C}^1(X, \mathbb{R})$, $S \subset V$, $c \in \mathbb{R}$, $\varepsilon, \delta > 0$ such that*

$$(\forall u \in \varphi^{-1}([c - 2\varepsilon, c + 2\varepsilon] \cap S_{2\delta})) : ||\varphi'(u)||_* \geq 8\varepsilon/\delta.$$

Then there exists $\eta \in \mathcal{C}([0, 1] \times V, V)$ such that
(i) $\eta(t, u) = u$, if $t = 0$ or if $u \notin \varphi^{-1}([c - 2\varepsilon, c + 2\varepsilon]) \cap S_{2\delta}$.
(ii) $\eta(1, \varphi^{c+\varepsilon} \cap S) \subset \varphi^{c-\varepsilon}$.
(iii) $\varphi(\eta(., u))$ is non increasing, $\forall u \in V$.

Proof. The proof is similar to the proof of Lemma 2.3. □

5.4 Minimax theorem

In this section, we assume that X, ψ and V satisfy assumption (A).

Let $\varphi \in \mathcal{C}^1(X, \mathbb{R})$ and Y be a closed subset of φ^d where $d \in \mathbb{R}$ is fixed. Define, for $j \geq 1$,

$$\mathcal{A}_j := \{A \subset \varphi^d : A \text{ is closed}, \ A \supset Y, \ \text{cat}_{\varphi^d, Y}(A) \geq j\},$$
$$c_j := \inf_{A \in \mathcal{A}_j} \sup_{u \in A} \varphi(u).$$

Theorem 5.16. *If*

$$(5.1) \qquad a := \sup_Y \varphi < c := c_k = \ldots = c_{k+m} \leq d,$$

then, for every $\varepsilon \in]0, (c-a)/2[$, $\delta > 0$, $A \in \mathcal{A}_{k+m}$ and $B \subset \varphi^d$ closed such that

$$\sup_A \varphi \leq c + \varepsilon, \quad \text{cat}_{\varphi^d}(B) \leq m,$$

there exists $u \in V$ such that
a) $c - 2\varepsilon \leq \varphi(u) \leq c + 2\varepsilon$,
b) $\text{dist}(u, A \backslash \overset{o}{B}) \leq 2\delta$,
c) $\|\varphi'(u)\|_* \leq 8\varepsilon/\delta$.

Proof. Suppose the thesis is false. We apply Lemma 5.15 with $S := A \backslash \overset{o}{B}$. We assume that

$$c - 2\varepsilon > a.$$

Hence we obtain

$$A \backslash \overset{o}{B} \prec_Y \varphi^{c-\varepsilon}.$$

It follows from Proposition 5.6 and the definition of c_k that

$$\begin{aligned} k + m \ &\leq \ \text{cat}_{\varphi^d, Y}(A) \\ &\leq \ \text{cat}_{\varphi^d, Y}(A \backslash \overset{o}{B}) + \text{cat}_{\varphi^d}(B) \\ &\leq \ \text{cat}_{\varphi^d, Y}(\varphi^{c-\varepsilon}) + m \\ &\leq \ k - 1 + m. \end{aligned}$$

This is a contradiction. $\square$

Definition 5.17. *The function $\varphi\big|_V$ satisfies the $(PS)_c$ condition if any sequence $(u_n) \subset V$ such that*

$$\varphi(u_n) \to c, \quad \|\varphi'(u_n)\|_* \to 0$$

has a convergent subsequence.

Theorem 5.18. *Under assumption (5.1), if $\varphi|_V$ satisfies the $(PS)_c$ condition, then $\operatorname{cat}_{\varphi^d}(K_c) \geq m + 1$.*

Proof. Assume that $\operatorname{cat}_{\varphi^d}(K_c) \leq m$. Proposition 5.9 implies the existence of a closed neighborhood B of K_c in φ^d such that $\operatorname{cat}_{\varphi^d}(B) \leq m$. By the preceding theorem, there exists a sequence $(u_n) \subset V$ satisfying

$$\varphi(u_n) \to c, \ \operatorname{dist}(u_n, \varphi^d \setminus \overset{o}{B}) \to 0, \|\varphi'(u_n)\|_* \to 0.$$

It follows from $(PS)_c$ condition that $K_c \cap (\varphi^d \setminus \overset{o}{B}) \neq \phi$, in contradiction with the definition of B. $\square$

Theorem 5.19. *If $\sup_Y \varphi < c_1$ and if $\varphi|_V$ satisfies the $(PS)_c$ condition for any $c \in [c_1, d]$, then $\varphi^{-1}([c_1, d])$ contains at least $\operatorname{cat}_{\varphi^d, Y}(\varphi^d)$ critical points of $\varphi|_V$.*

Proof. If $n := \operatorname{cat}_{\varphi^d, Y}(\varphi^d)$, we obtain

$$\sup_Y \varphi < c_1 \leq c_2 \leq \ldots \leq c_n \leq d.$$

It suffices then to use the preceding theorem. $\square$

Theorem 5.20. *If $\varphi|_V$ is bounded from below and satisfies the $(PS)_c$ condition for any $c \in [\inf_V \varphi, d]$, then $\varphi|_V$ has a minimum and φ^d contains at least $\operatorname{cat}_{\varphi^d}(\varphi^d)$ critical points of $\varphi|_V$.*

Proof. By the preceding theorem applied to $Y = \phi$, φ^d contains at least $\operatorname{cat}_{\varphi^d}(\varphi^d)$ critical points of $\varphi|_V$. By Theorem 5.18, $c_1 = \inf_V \varphi$ is a critical value of $\varphi|_V$. $\square$

5.5   Critical nonlinearities

In this section, we consider the problem

$$(\mathcal{P}_\lambda) \qquad \begin{cases} -\Delta u + \lambda u = |u|^{2^* - 2} u, \\ u \geq 0, u \in H_0^1(\Omega), \end{cases}$$

where Ω is a smooth bounded domain of $\mathbb{R}^N$, $N \geq 4$ and $\lambda > -\lambda_1(\Omega)$. By Theorem 1.45, problem $(\mathcal{P}_\lambda)$ has a nontrivial solution if $-\lambda_1(\Omega) < \lambda < 0$. We shall prove the existence of $-\lambda_1(\Omega) < \lambda^* < 0$ such that, for $\lambda^* < \lambda < 0$, problem $(\mathcal{P}_\lambda)$ has at least $\operatorname{cat}_\Omega(\Omega)$ nontrivial solutions.

By Corollary 1.13, the functional

$$\psi(u) := \int_\Omega (u^+)^{2^*} dx$$

is of class $C^2(H_0^1(\Omega), \mathbb{R})$. We define the quadratic functional

$$\varphi_\lambda(u) := \int_\Omega [|\nabla u|^2 + \lambda u^2] dx$$

on the manifold

$$V := \{u \in H_0^1(\Omega) : \psi(u) = 1\}.$$

On $H_0^1(\Omega)$, we choose the norm $||u|| := |\nabla u|_2$.

Lemma 5.21. *Any sequence $(u_n) \subset V$ such that*

$$\varphi_\lambda(u_n) \to c < S, \quad ||\varphi_\lambda'(u_n)||_* \to 0,$$

contains a convergent subsequence.

Proof. By assumption, there exists a sequence $(\mu_n) \subset \mathbb{R}$ such that

$$-\Delta u_n + \lambda u_n - \mu_n (u_n^+)^{2^*-1} \to 0 \text{ in } H^{-1}(\Omega).$$

Therefore $\varphi_\lambda(u_n) - \mu_n \to 0$ and $\mu_n \to c$. If we define $v_n := \mu_n^{(N-2)/4} u_n$, we obtain

$$\int_\Omega \left[\frac{|\nabla v_n|^2}{2} + \frac{\lambda v_n^2}{2} - \frac{(v_n^+)^{2^*}}{2^*} \right] dx \to c^{N/2} N$$

and

$$-\Delta v_n + \lambda v_n - (v_n^+)^{2^*-1} \to 0 \text{ in } H^{-1}(\Omega).$$

By Lemma 1.44, the sequence (v_n) contains a convergent subsequence. But then (u_n) contains also a convergent subsequence. □

Lemma 5.22. *If $N \geq 4$ and $-\lambda_1(\Omega) < \lambda < 0$, then*

$$m(\lambda, \Omega) := \inf_{u \in V} \varphi_\lambda(u) < S$$

and there exists $u \in V$ such that $\varphi_\lambda(u) = m(\lambda, \Omega)$.

Proof. By Lemma 1.46, we have $m(\lambda, \Omega) < S$. Theorem 5.20 and the preceding lemma imply the existence of $u \in V$ such that $\varphi_\lambda(u) = m(\lambda, \Omega)$. □

On V, we define the map

$$\beta(u) := \int_\Omega (u^+)^{2^*} x \, dx.$$

Lemma 5.23. *If $(u_n) \subset V$ is such that $||u_n||^2 \to S$, then $\text{dist}(\beta(u_n), \Omega) \to 0$.*

Proof. Assume, by contradiction, that $\text{dist}(\beta(u_n), \Omega) \not\to 0$. By going if necessary to a subsequence, we can assume that

$$\text{dist}(\beta(u_n), \Omega) > r > 0,$$
$$u_n^+ \rightharpoonup u \qquad\qquad \text{in } D^{1/2}(\mathbb{R}^N),$$
$$|\nabla(u_n^+ - u)|^2 \rightharpoonup \mu \qquad \text{in } \mathcal{M}(\mathbb{R}^N),$$
$$|u_n^+ - u|^{2^*} \rightharpoonup \nu \qquad \text{in } \mathcal{M}(\mathbb{R}^N),$$
$$u_n^+ \to u \qquad\qquad \text{a.e. on } \Omega.$$

Since Ω is bounded, Lemma 1.40 implies that

$$S = |\nabla u|_2^2 + \|\mu\|, \quad 1 = |u|_{2^*}^{2^*} + \|\nu\|,$$

and

$$\|\nu\|^{2/2^*} \le S^{-1}\|\mu\|, \quad |u|_{2^*}^2 \le S^{-1}|\nabla u|_2^2.$$

It follows that $|u|_{2^*}^{2^*}$ and $\|\nu\|$ are equal either to 0 or to 1. By Proposition 1.43, $u = 0$. We deduce from Lemma 1.40 that ν is concentrated at a single point y of $\bar{\Omega}$ and so

$$\beta(u_n) \to \int_\Omega x\, d\nu(x) = y \in \bar{\Omega}.$$

This is a contradiction. $\square$

Since Ω is a smooth bounded domain of $\mathbb{R}^N$, we choose $r > 0$ small enough that

$$\Omega_r^+ := \{x \in \mathbb{R}^N : \text{dist}(x, \Omega) < r\}$$

and

$$\Omega_r^- := \{x \in \Omega : \text{dist}(x, \partial\Omega) > r\}$$

are homotopically equivalent to Ω. Moreover we can assume that $B[0, r] \subset \Omega$. We define

$$m(\lambda) := m(\lambda, B(0, r)) < S.$$

Lemma 5.24. *There exists* $-\lambda_1(\Omega) < \lambda^* < 0$ *such that, for* $\lambda^* < \lambda < 0$,

$$u \in \varphi_\lambda^{m(\lambda)} \Rightarrow \beta(u) \in \Omega_r^+.$$

Proof. By the Hölder inequality, for every $u \in V$

$$|u|_2^2 \le |\Omega|^{2/N}|u|_{2^*}^2 = |\Omega|^{2/N}.$$

The preceding lemma implies the existence of $\varepsilon > 0$ such that

$$u \in V, \|u\|^2 \le S + \varepsilon \Rightarrow \beta(u) \in \Omega_r^+.$$

We choose $\lambda^* := -\varepsilon/|\Omega|^{2/N}$. If $\lambda^* < \lambda < 0$ and $u \in \varphi_\lambda^{m(\lambda)}$, we obtain

$$\|u\|^2 \le m(\lambda) - \lambda|u|_2^2 \le S - \lambda^*|\Omega|^{2/N} = S + \varepsilon,$$

so that $\beta(u) \in \Omega_r^+$. $\square$

Lemma 5.25. *If $N \geq 4$ and $\lambda^* < \lambda < 0$, then $\operatorname{cat}_{\varphi_\lambda^{m(\lambda)}}(\varphi_\lambda^{m(\lambda)}) \geq \operatorname{cat}_\Omega(\Omega)$.*

Proof. Define $\gamma : \Omega_r^- \to \varphi_\lambda^{m(\lambda)}$ by

$$\gamma(y)(x) := v(x - y), \quad x \in B(y, r),$$
$$:= 0, \qquad\qquad x \notin B(y, r),$$

where $v \in H_0^1(B(0, r))$ is such that $v \geq 0$, $|v|_{2^*} = 1$ and

$$\int_{B(0,r)} [|\nabla v|^2 + \lambda v^2] dx = m(\lambda).$$

Since v is a radial function, it is clear that $\beta \circ \gamma = \operatorname{id}$. Assume that

$$\varphi_\lambda^{m(\lambda)} = A_1 \cup \ldots \cup A_n,$$

where A_j, $j = 1, ..., n$, is closed and contractible in $\varphi_\lambda^{m(\lambda)}$, i.e. there exists $h_j \in \mathcal{C}([0, 1] \times A_j, \varphi_\lambda^{m(\lambda)})$ such that, for every $u, v \in A_j$,

$$h_j(0, u) = u, \quad h_j(1, u) = h_j(1, v).$$

Consider $B_j := \gamma^{-1}(A_j)$, $1 \leq j \leq n$. The sets B_j are closed and

$$\Omega_r^- = B_1 \cup \ldots \cup B_n.$$

Using the deformation

$$g_j(t, x) := \beta\Big(h_j(t, \gamma(x))\Big),$$

we prove that B_j is contractible in Ω_r^+ by the preceding lemma. It follows that

$$\operatorname{cat}_\Omega(\Omega) = \operatorname{cat}_{\Omega_r^+}(\Omega_r^-) \leq n. \qquad \square$$

Theorem 5.26. *If Ω is a smooth bounded domain of $\mathbb{R}^N$, $N \geq 4$, there exists $-\lambda_1(\Omega) < \lambda^* < 0$ such that, for $\lambda^* < \lambda < 0$, problem $(\mathcal{P}_\lambda)$ has at least $\operatorname{cat}_\Omega(\Omega)$ nontrivial solutions.*

Proof. By Lemmas 5.21 and 5.22, for $c \leq m(\lambda) \leq m(\lambda, \Omega) < S$, φ_λ satisfies the $(PS)_c$ condition. Theorem 5.20 and Lemma 5.25 imply that, for $\lambda^* < \lambda < 0$, $\varphi_\lambda^{m(\lambda)}$ contains at least $n := \operatorname{cat}_\Omega(\Omega)$ critical points of $\varphi_\lambda|_V$, $u_1, ..., u_n$.

For $j = 1, ..., n$, there exists $\mu_j \in \mathbb{R}$ such that

$$-\Delta u_j + \lambda u_j - \mu_j (u_j^+)^{2^*-1} = 0.$$

Multiplying the equation by u_j^- and integrating over Ω, we find

$$0 = |\nabla u_j^-|_2^2 + \lambda |u_j^-|_2^2.$$

Since $-\lambda_1(\Omega) < \lambda$, $u_j^- = 0$ and

$$-\Delta u_j + \lambda u_j - \mu_j u_j^{2^*-1} = 0.$$

It follows that $\mu_j = \varphi_\lambda(u_j)$ and $v_j := \mu_j^{(N-2)/4} u_j$ is a nontrivial solution of $(\mathcal{P}_\lambda)$. $\square$

The preceding theorem is due to Rey when $N \geq 5$ and to Lazzo when $N = 4$.

Chapter 6

Generalized linking theorem

6.1 Degree theory

This section is devoted to the degree theory of Kryszewski and Szulkin.

Let (e_k) be a total orthonormal sequence in a separable Hilbert space E and define

$$(6.1) \qquad |||u||| := \sum_{k=0}^{\infty} \frac{1}{2^{k+1}} |(u, e_k)|.$$

The topology generated by $|||.|||$ will be denoted by σ and all topological notions related to it will include this symbol. It is clear that $|||u||| \leq ||u||$.

Remarks 6.1. a) If (u_n) is bounded in E, then

$$u_n \rightharpoonup u \Longleftrightarrow u_n \xrightarrow{\sigma} u.$$

b) The space E with the norm $|||.|||$ is not complete.

Definition 6.2. Let U be an open bounded subset of E such that $\bar{U}$ is σ-closed. A map $f : \bar{U} \to E$ is admissible if
a) $0 \notin f(\partial U)$,
b) f is σ-continuous,
c) each point $u \in U$ has a σ-neighborhood N_u such that $(\mathrm{id} - f)(N_u \cap U)$ is contained in a finite-dimensional subspace of E.

Definition 6.3. Let $f : \bar{U} \to E$ be admissible. By assumption $f^{-1}(0)$ is a σ-closed subset of U. Hence $f^{-1}(0)$ is σ-compact. For every $u \in f^{-1}(0)$, let N_u be a σ-open neighborhood of u such that $(\mathrm{id} - f)(N_u \cap U)$ is contained in a finite-dimensional subspace of E. There are points $u_1, ..., u_m \in f^{-1}(0)$ such that $f^{-1}(0) \subset V := \cup_{n=1}^{m} N_{u_n} \cap U$. The set V is open and there exists a finite-dimensional subspace F of E such that $(\mathrm{id} - f)(V) \subset F$. The degree is defined by

$$\deg(f, U) := \deg_B(f|_{V \cap F}, V \cap F),$$

where $\deg_B$ is the Brouwer degree.

Proposition 6.4. *The degree of admissible maps is well defined.*

Proof. a) Let us define $\tilde{f} = f|_{V \cap F}$. It is clear that $\tilde{f}^{-1}(0) = f^{-1}(0) \subset V \cap F$ is compact and that $V \cap F$ is open in F. Moreover $\tilde{f}$ is continuous.

b) Let G be another finite-dimensional subspace of E such that $(\mathrm{id} - f)(V) \subset G$. We may assume that $F \subset G$. By the contraction property of the Brouwer degree (Theorem D.20), we obtain

$$\deg_B(f|_{V \cap F}, V \cap F) = \deg_B(f|_{V \cap G}, V \cap G).$$

Thus the degree does not depend on the choice of F.

c) Let W be another neighborhood of $f^{-1}(0)$ such that $(\mathrm{id} - f)(W) \subset F$. We may assume that $V \subset W$. The excision property of the Brouwer degree implies that

$$\deg_B(f|_{V \cap F}, V \cap F) = \deg_B(f|_{W \cap F}, W \cap F).$$

Thus the degree does not depend on the choice of V. $\square$

Definition 6.5. A map $h : [0, 1] \times \bar{U} \to E$ is an admissible homotopy if
a) $0 \notin h([0, 1] \times \partial U)$,
b) h is σ-continuous,
c) each point $(t, u) \in [0, 1] \times U$ has a σ-neighborhood $N_{(t,u)}$ such that

$$\{v - h(s, v) : (s, v) \in N_{(t,u)} \cap ([0, 1] \times U)\}$$

is contained in a finite-dimensional subspace of E.

Theorem 6.6. a) (*Normalization*). If $v \in U$ then $\deg(\mathrm{id} - v, U) = 1$.
b) (*Existence*). If f is admissible and if $\deg(f, U) \neq 0$ then $0 \in f(U)$.
c) (*Homotopy invariance*).   If h is an admissible homotopy then $\deg(h(t, .), U)$ is independent of $t \in [0, 1]$.

Proof. a) Normalization is trivial.

b) Existence follows from the corresponding property of the Brouwer degree.

c) Since $h^{-1}(0)$ is σ-compact, the projection K of $h^{-1}(0)$ into U is also σ-compact. Thus there exists an open subset W of $[0,1] \times U$ such that $[0,1] \times K \subset W$ and

$$\{u - h(t,u) : (t,u) \in W\}$$

is contained in a finite-dimensional subspace F of E. But then K is compact and there exists an open subset V of U such that $[0,1] \times K \subset [0,1] \times V \subset W$. Since, by definition,

$$\deg(h(t,.),U) = \deg_B(h(t,.)\big|_{V \cap F}, V \cap F),$$

the homotopy invariance follows from the corresponding property of the Brouwer degree. $\square$

6.2 Pseudogradient flow

Let Y be a separable subspace of a Hilbert space X and let $Z := Y^\perp$. Let $P : X \to Y$, $Q : X \to Z$ be the orthogonal projections. On X we define the norm

$$(6.2) \qquad |||u||| := \max\left(||Qu||, \sum_{k=1}^\infty \frac{1}{2^{k+1}}|(Pu, e_k)|\right)$$

where (e_k) is a total orthonormal sequence in Y. The topology generated by $|||.|||$ will be denoted by τ and all topological notions related to it will include this symbol. It is clear that

$$||Qu|| \le |||u||| \le ||u||.$$

Our basic assumption is

(A) $\varphi \in C^1(X, \mathbb{R})$ is τ-upper semicontinuous, $\nabla\varphi$ is weakly sequentially continuous and there exists $\alpha < \beta$ and $\varepsilon > 0$ such that

$$\forall u \in \varphi^{-1}([\alpha, \beta]) : ||\varphi'(u)|| \ge \varepsilon.$$

The following lemmas are due to Kryszewski and Szulkin.

Lemma 6.7. *Under assumption (A), there exists a τ-open neighborhood V of φ^β and a vector field $f : V \to X$ satisfying:*
a) f is locally Lipschitz continuous and τ-locally Lipschitz continuous,
b) each point $u \in V$ has a τ-neighborhood V_u such that $f(V_u)$ is contained in a finite-dimensional subspace of X,
c) $m := \sup_{u \in V} ||f(u)|| < \infty$, $\forall u \in V$, $(\nabla\varphi(u), f(u)) \ge 0$,
d) $\forall u \in \varphi^{-1}([\alpha, \beta])$, $(\nabla\varphi(u), f(u)) > 1$.

Proof. Let us define, on $\varphi^{-1}([\alpha, \beta])$,

$$g(v) := \frac{2\|\nabla\varphi(v)\|}{\|\nabla\varphi(v)\|^2}.$$

Since $\nabla\varphi$ is weakly sequentially continuous, there exists a τ-open neighborhood N_v of v such that

$$(\nabla\varphi(u), g(v)) > 1$$

for every $u \in N_v$. Since φ is τ-upper semicontinuous, $\tilde{N} := \varphi^{-1}(]-\infty, \alpha[)$ is τ-open. The family

$$\mathcal{N} := \{N_v : \alpha \le \varphi(v) \le \beta\} \cup \{\tilde{N}\}$$

is a τ-open covering of φ^β. Since (φ^β, τ) is metric, hence paracompact, there exists a τ-locally finite τ-open covering $\mathcal{M} := \{M_i : i \in I\}$ of φ^β finer than $\mathcal{N}$. We define the τ-open neighborhood of φ^β:

$$V := \bigcup_{i \in I} M_i.$$

If $M_i \subset N_v$ for some $v \in \varphi^{-1}([\alpha, \beta])$, we choose $v_i := v$. If $M_i \subset \tilde{N}$, we choose $v_i := 0$. Let $\{\lambda_i : i \in I\}$ be a τ-Lipschitz continuous partition of unity subordinated to $\mathcal{M}$ and define on V

$$f(u) := \sum_{i \in I} \lambda_i(u)v_i.$$

It is easy to verify the properties of the lemma. $\square$

We now consider the Cauchy problem

$$\frac{d}{dt}\eta(t, u) = -f(\eta(t, u)),$$
$$\eta(0, u) = u \in \varphi^\beta.$$

Lemma 6.8. *Under assumption (A), the flow η is well defined on $\mathbb{R}^+ \times \varphi^\beta$ and satisfies the following properties:*
a) there exists $T > 0$ such that

$$\eta(T, \varphi^\beta) \subset \varphi^\alpha,$$

b) each point $(t, u) \in [0, T] \times \varphi^\beta$ has a τ-neighborhood $N_{(t,u)}$ such that

$$\{v - \eta(s, v) : (s, v) \in N_{(t,u)} \cap ([0, T] \times \varphi^\beta)\}$$

is contained in a finite-dimensional subspace of X,
c) η is τ-continuous.

Proof. a) Since

$$\frac{d}{dt}\varphi(\eta(t,u)) = \left(\nabla\varphi(\eta(t,u)), \frac{d}{dt}\eta(t,u)\right)$$
$$= -(\nabla\varphi(\eta(t,u)), f(\eta(t,u))) \le 0,$$

$\varphi(\eta(.,u))$ is nonincreasing and φ^β is positively invariant. But f is bounded on V, so that η is well defined on $\mathbb{R}^+ \times \varphi^\beta$. Let $u \in \varphi^\beta$. We choose $T := (\beta - a)$. If there exists $t \in [0,T]$ such that $\varphi(\eta(t,u)) < \alpha$, then $\varphi(\eta(T,u)) < \alpha$. If

$$\eta(t,u) \in \varphi^{-1}([\alpha,\beta]), \forall t \in [0,T],$$

we obtain

$$\varphi(\eta(T,u)) = \varphi(u) + \int_0^T \frac{d}{dt}\varphi(\eta(t,u))dt$$
$$= \varphi(u) - \int_0^T (\nabla\varphi(\eta(t,u)), f(\eta(t,u))) \, dt$$
$$\le \beta - (\beta - \alpha) = \alpha.$$

Thus property a) is satisfied.

b) Let $(t_0, u_0) \in [0,T] \times \varphi^\beta$. The set $K := \eta([0,T] \times \{u_0\})$ is τ-compact. Thus there exists $r > 0$ and $k > 0$ such that

$$U := \{u \in X : |||u - K||| < r\} \subset V,$$
$$u, v \in U \Rightarrow |||f(u) - f(v)||| \le k|||u - v|||$$

and $f(U)$ is contained in a finite-dimensional subspace W of X. If $\eta(s,u) \in U$ for $0 \le s \le t \le T$, we obtain

$$|||\eta(t,u) - \eta(t,u_0)||| \le |||u - u_0||| + \int_0^t |||f(\eta(s,u)) - f(\eta(s,u_0))|||ds$$
$$\le |||u - u_0||| + k \int_0^t |||\eta(s,u) - \eta(s,u_0)|||ds.$$

The next lemma implies that

$$|||\eta(t,u) - \eta(t,u_0)||| \le |||u - u_0|||e^{kt} \le |||u - u_0|||e^{kT}.$$

In particular for $|||u - u_0||| < re^{-kT}$ and $0 < t < T$, we obtain

$$u - \eta(t,u) = \int_0^t f(\eta(s,u(s)))ds \in W.$$

c) Let $0 < \delta < re^{-kT}$. If $|||u - u_0||| < \delta$, $|t - t_0| < \delta$ and $0 < t < T$, we have

$$|||\eta(t,u) - \eta(t_0,u_0)||| \le |||\eta(t,u) - \eta(t,u_0)||| + \int_{t_0}^t |||f(\eta(s,u_0))|||ds$$
$$\le (e^{kT} + m)\delta.$$

Thus η is τ-continuous. $\square$

Lemma 6.9. (Gronwall inequality). *If $a, b \geq 0$ and $f \in C([0,T], \mathbb{R}^+)$ satisfy*

$$\forall t \in [0,T], \quad f(t) \leq a + b \int_0^t f(s)ds$$

then it follows that

$$\forall t \in [0,T], \quad f(t) \leq ae^{bt}.$$

Proof. If $b = 0$ the result is obvious. Assuming $b > 0$, we obtain

$$\frac{d}{dt}\left(e^{-bt}\int_0^t f(s)ds\right) \leq ae^{-bt}$$

and so

$$e^{-bt}\int_0^t f(s)ds \leq \frac{a}{b}(1 - e^{-bt}).$$

It follows that

$$f(t) \leq a + b\left[\frac{a}{b}(e^{bt} - 1)\right] = ae^{bt}. \qquad \square$$

6.3   Generalized linking theorem

Let Y be a separable subspace of a Hilbert space X and let $Z = Y^{\perp}$. Let $P : X \to Y$, $Q : X \to Z$ be the orthogonal projections. The τ-topology on X is generated by the norm (6.2). Let $\rho > r > 0$ and let $z \in Z$ be such that $||z|| = 1$. Define

$$
\begin{aligned}
M &:= \{u = y + \lambda z : ||u|| \leq \rho, \lambda \geq 0, y \in Y\}, \\
M_0 &:= \{u = y + \lambda z : y \in Y, ||u|| = \rho \text{ and } \lambda \geq 0 \text{ or } ||u|| \leq \rho \text{ and } \lambda = 0\}, \\
N &:= \{u \in Z : ||u|| = r\}.
\end{aligned}
$$

Let $\varphi \in C^1(X, \mathbb{R})$ be such that

(6.3) φ is τ-upper semicontinuous and $\nabla\varphi$ is weakly sequentially continuous,

(6.4) $b := \inf_N \varphi > 0 = \sup_{M_0} \varphi, \quad d := \sup_M \varphi < \infty.$

Theorem 6.10. (Kryszewski-Szulkin, 1996). *If φ satisfies (6.3) and (6.4), there exists $c \in [b, d]$ and a sequence $(u_n) \subset X$ such that*

$$\varphi(u_n) \to c, \quad \varphi'(u_n) \to 0.$$

Proof. a) If the conclusion of the theorem is not satisfied, there exists $\varepsilon > 0$ such that

$$\forall u \in \varphi^{-1}([b - \varepsilon, d + \varepsilon]) : ||\varphi'(u)|| \geq \varepsilon.$$

By (6.3), assumption (A) is satisfied with $\alpha := b - \varepsilon$ and $\beta := d + \varepsilon$. Let η be the flow given by Lemma 6.8. Since $M \subset \varphi^{d+\varepsilon}$, there exists $T > 0$ such that

$$(6.5) \qquad \eta(T, M) \subset \varphi^{b-\varepsilon}.$$

b) Let U be the interior of M in $E := Y \oplus \mathbb{R}z$. The topology induced on E by τ is the σ-topology generated by the norm (6.1). We define the homotopy $h : [0, T] \times M \to E$ by

$$h(t, u) := (P\eta(t, u), (\|Q\eta(t, u)\| - r)z).$$

It is clear that

$$(6.6) \qquad h(t, u) = 0 \iff \eta(t, u) \in N.$$

By assumption (6.4), $\inf_N \varphi = b > 0$ and $\sup_{\partial U} \varphi = 0$, so that

$$0 \notin h([0, T] \times \partial U).$$

Lemma 6.8 implies that h is σ-continuous and that each point $(t, u) \in [0, T] \times M$ has a σ-neighborhood $N_{(t,u)}$ such that

$$\{v - h(s, v) : (s, v) \in N_{(t,u)} \cap ([0, T] \times U)\}$$

is contained in a finite-dimensional subspace of E. Thus the homotopy h is admissible. Using normalization and homotopy invariance, we obtain

$$\deg(h(T, .), U) = \deg(h(0, .), U) = \deg(\mathrm{id} - rz, U) = 1.$$

By existence, there is $u \in U$ such that $h(T, u) = 0$. It follows from (6.6) that $\eta(T, u) \in N$ and so $\varphi(\eta(T, u)) \geq b$. On the other hand, by (6.5), $\varphi(\eta(T, u)) \leq b - \varepsilon$. This is a contradiction. $\square$

The following result is left to the reader.

Corollary 6.11. Let $\psi \in \mathcal{C}^1(X, \mathbb{R})$ be weakly sequentially lower semi-continuous, bounded below and such that $\nabla \psi$ is weakly sequentially continuous. If

$$\varphi(u) := \frac{\|Qu\|^2}{2} - \frac{\|Pu\|^2}{2} - \psi(u)$$

satisfies (6.4) then the conclusion of Theorem 6.10 holds. $\square$

6.4   Semilinear Schrödinger equation

In this section, we consider the problem

$$(\mathcal{P}) \qquad \begin{cases} -\Delta u + V(x)u = f(x, u), \\ u \in H^1(\mathbb{R}^N). \end{cases}$$

We assume that

$$F(x, u) := \int_0^u f(x, s)ds$$

is subquadratic at the origin and superquadratic but subcritical at infinity. We will prove the existence of a nontrivial solution by applying the generalized linking theorem to the functional

$$\varphi(u) := \int_{\mathbb{R}^N} \Big[\frac{|\nabla u|^2}{2} + V(x)\frac{u^2}{2} - F(x, u)\Big]dx,$$

defined on $H^1(\mathbb{R}^N)$. We assume that V and f are periodic with respect to x and that 0 lies in a gap of the spectrum of $-\Delta + V$. Because of the periodicity with respect to x, φ is invariant under a $\mathbb{Z}^N$ action and the Palais-Smale condition fails at every critical level. Since 0 lies in a gap of the essential spectrum, $H^1(\mathbb{R}^N)$ is the sum of two infinite dimensional subspaces on which the quadratic part of φ is negative and positive respectively.

We assume that

(f_1) $V \in \mathcal{C}(\mathbb{R}^N)$ and $f \in \mathcal{C}(\mathbb{R}^N \times \mathbb{R})$ are 1-periodic in x_k, $1 \le k \le$ and the linear operator

$$L : H^2(\mathbb{R}^N) \to L^2(\mathbb{R}^N) : u \mapsto -\Delta u + Vu$$

is invertible.

We denote by $|L|^{1/2}$ the square root of the absolute value of L. The domain of $|L|^{1/2}$ is the space $X := H^1(\mathbb{R}^N)$. On X, we choose the inner product

$$a(u, v) := \int_{\mathbb{R}^N} |L|^{1/2}u|L|^{1/2}v\,dx$$

and the corresponding norm

$$\|u\|_a := \sqrt{a(u, u)}.$$

There exists an orthogonal decomposition $X = Y \oplus Z$ such that

$$\forall u \in Y, \quad \int [|\nabla u|^2 + Vu^2]dx = -a(u, u),$$

$$\forall u \in Z, \quad \int [|\nabla u|^2 + Vu^2]dx = a(u, u).$$

We denote by P (resp. Q) the orthogonal projector onto Y (resp. Z). The spaces Y and Z are also orthogonal with respect to the L^2 inner product.

We assume also that

(f_2) For some $2 < q < p < 2^*$, $c > 0$

$$|f(x, u)| \leq c(|u|^{q-1} + |u|^{p-1}).$$

(f_3) There exists $\alpha > 2$ such that, for every $u \neq 0$,

$$0 < \alpha F(x, u) \leq u f(x, u).$$

Lemma 6.12. *Under assumptions* (f_1)-(f_2), $\varphi \in \mathcal{C}^1(X, \mathbb{R})$.

Proof. It suffices to use Lemma 3.10. $\square$

Lemma 6.13. *Under assumptions* (f_1)-(f_2), *there exists* $r > 0$ *such that*

$$b := \inf_{\substack{u \in Z \\ ||u||_a = r}} \varphi(u) > 0 = \min_{\substack{u \in Z \\ ||u||_a \leq r}} \varphi(u).$$

Proof. By (f_2), there exists $c_1 > 0$ such that

$$|F(x, u)| \leq c_1(|u|^q + |u|^p).$$

We obtain on Z,

$$\varphi(u) \geq \frac{||u||_a^2}{2} - c_1 \int (|u|^q + |u|^p)$$

$$= \frac{||u||_a^2}{2} - c_1(|u|_q^q + |u|_p^p).$$

It suffices then to use the Sobolev imbedding Theorem. $\square$

Lemma 6.14. *Under assumptions* (f_1)-(f_3), *there exists* $z \in Z$ *and* $\rho > r$ *such that* $||z||_a = r$ *and*

$$\max_{M_0} \varphi = 0, \quad \sup_M \varphi < \infty,$$

where

$$M := \{u = y + \lambda z : ||u||_a < \rho, \lambda \geq 0, y \in Y\},$$
$$M_0 := \{u = y + \lambda z : y \in Y, ||u||_a = \rho \text{ and } \lambda \geq 0 \text{ or } ||u||_a \leq \rho \text{ and } \lambda = 0\}.$$

Proof. 1) By assumption (f_3), we have, on Y,

$$\varphi(u) \leq -||u||_a^2 - \int F(x, u) dx \leq 0.$$

2) Assumptions (f_2) and (f_3) imply the existence of $c_2 > 0$ such that

$$c_2|u|^\alpha - \frac{c_3}{4}|u|^2 \le F(x, u)$$

where

$$c_3 := \inf_{\substack{u \in X \\ |u|_a = 1}} \|u\|_a^2.$$

We choose $z \in Z$ such that $\|z\|_a = r$ and we denote by E the closure of $Y \oplus \mathbb{R}z$ in $L^\alpha(\mathbb{R}^N)$. Since there exists a continuous projection $E \to \mathbb{R}z$, we have, for some $c_4 > 0$ and for every $\lambda \ge 0$ and $y \in Y$,

$$\varphi(y + \lambda z) \le -\frac{\|y\|_a^2}{2} + \frac{\lambda^2}{2}\|z\|_a^2 + \frac{c_3}{4}|y + \lambda z|_2^2 - c_2|y + \lambda z|_\alpha^\alpha$$

$$\le -\frac{\|y\|_a^2}{4} + c_4(\lambda^2 - \lambda^\alpha).$$

It follows that

$$\varphi(u) \to -\infty, \quad \|u\|_a \to \infty, \quad u \in Y \oplus \mathbb{R}^+ z,$$

and so, for some $\rho > r$, $0 = \max_{M_0} \varphi$.

3) By (f_2), φ maps bounded sets into bounded sets, hence $\sup_M \varphi < \infty$. $\square$

Lemma 6.15. *Under assumptions (f_1)-(f_3), there exists $c \in [b, d]$ and a sequence $(u_n) \subset X$ such that*

$$(6.7) \qquad \varphi(u_n) \to c, \quad \varphi'(u_n) \to 0.$$

Proof. a) Let us prove that φ is τ-upper semicontinuous. Assume that $u_n \overset{\tau}{\longrightarrow} u$ and $c \le \varphi(u_n)$. Since $Qu_n \to Qu$ and F is nonnegative, (Pu_n) is bounded so that $Pu_n \rightharpoonup Pu$. It follows that $u_n \to u$ in $L^2_{loc}(\mathbb{R}^N)$ and, going if necessary to a subsequence, $u_n \to u$ a.e. on $\mathbb{R}^N$. Using the Fatou lemma, we obtain

$$-\varphi(u) = \frac{\|Pu\|_a^2}{2} - \frac{\|Qu\|_a^2}{2} + \int F(x, u)dx$$

$$\le \underline{\lim}\left[\frac{\|Pu_n\|_a^2}{2} - \frac{\|Qu_n\|_a^2}{2} + \int F(x, u_n)dx\right] = \underline{\lim} -\varphi(u_n) \le -c.$$

b) Let us prove that $\nabla\varphi$ is weakly sequentially continuous. Assume that $u_n \rightharpoonup u$ in $H^1(\mathbb{R}^N)$. Since $u_n \to u$ in $L^2_{loc}(\mathbb{R}^N)$, for every $w \in \mathcal{D}(\mathbb{R}^N)$, we have

$$(\nabla\varphi(u), w) = \lim_{n \to \infty}(\nabla\varphi(u_n), w).$$

By (f_2), $(\nabla\varphi(u_n))$ is bounded in $H^1(\mathbb{R}^N)$, so that $\nabla\varphi(u_n) \rightharpoonup \nabla\varphi(u)$ in $H^1(\mathbb{R}^N)$.

c) It suffices then to use Theorem 6.10 and Lemmas 6.12, 6.13 and 6.14. $\square$

Lemma 6.16. *Under assumptions* (f$_1$)-(f$_3$), *any sequence satisfying* (6.7) *is bounded in* $H^1(\mathbb{R}^N)$.

Proof. Let $\theta \in C^\infty(\mathbb{R}, \mathbb{R})$ be such that, $0 \leq \theta \leq 1$, $\theta(t) = 1$, $|t| \geq 2$, $\theta(t) = 0$, $|t| \leq 1$. Define

$$g(x, u) := \theta(u)f(x, u), \quad h(x, u) := (1 - \theta(u))f(x, u),$$
$$r := p/(p - 1), \qquad s := q/(q - 1).$$

By assumption, we have

$$d|g(x, u)|^r \leq ug(x, u),$$
$$d|h(x, u)|^s \leq uh(x, u).$$

For n big enough, we obtain from (6.7)

$$(6.8) \qquad c + 1 + \|u\|_a \geq \varphi(u_n) - \frac{1}{2}\langle \varphi'(u_n), u_n \rangle$$
$$= \int [\frac{1}{2}u_n f(x, u_n) - F(x, u_n)]dx$$
$$\geq (\frac{1}{2} - \frac{1}{\alpha}) \int u_n f(x, u_n)dx$$
$$\geq d(\frac{1}{2} - \frac{1}{\alpha})[|g(x, u_n)|_r^r + |h(x, u_n)|_s^s].$$

Let us write $u_n = y_n + z_n$, $y_n \in Y$, $z_n \in Z$. We deduce from (6.7) and (6.8) that, for n large enough and some $c_7 > 0$,

$$a(y_n, y_n) = -\langle \varphi'(u_n), y_n \rangle - \int y_n f(x, u_n)dx$$
$$\leq \|y_n\|_a + [|y_n|_p|g(x, u_n)|_r + |y_n|_q|h(x, u_n)|_s]$$
$$\leq \|y_n\|_a + c_7\|y_n\|_a[(1 + \|u_n\|_a)^{1/r} + (1 + \|u_n\|_a)^{1/s}]$$

and, similarly,

$$a(z_n, z_n) \leq \|z_n\|_a + c_7\|z_n\|_a[(1 + \|u_n\|_a)^{1/r} + (1 + \|u_n\|_a)^{1/s}].$$

It is then easy to verify that $\|u_n\|_a^2 = a(y_n, y_n) + a(z_n, z_n)$ is bounded. $\square$

Theorem 6.17. (Kryszewski-Szulkin, 1996). *Under assumptions* (f$_1$)-(f$_3$), *problem* ($\mathcal{P}$) *has a nontrivial solution.*

Proof. Lemma 6.15 implies the existence of a sequence $(u_n) \subset H^1(\mathbb{R}^N)$ satisfying

$$\varphi(u_n) \to c > 0, \quad \varphi'(u_n) \to 0.$$

By the preceding lemma, (u_n) is bounded in $H^1(\mathbb{R}^N)$.

If

$$\delta := \varlimsup_{n\to\infty} \sup_{y\in\mathbb{R}^N} \int_{B(y,1)} |u_n|^2 = 0$$

then, by Lemma 1.21, $u_n \to 0$ in $L^s(\mathbb{R}^N)$ for $2 < s < 2^*$. It follows that

$$0 < c = \varphi(u_n) - \frac{1}{2}\langle\varphi'(u_n), u_n\rangle + o(1)$$

$$= \int [\frac{1}{2}u_n f(x, u_n) - F(x, u_n)]dx + o(1) = o(1).$$

This is a contradiction. Thus $\delta > 0$.

Going if necessary to a subsequence, we may assume the existence of $k_n \in \mathbb{Z}^N$ such that

$$\int_{B(k_n, 1+\sqrt{N})} |u_n|^2 > \delta/2.$$

Let us define $v_n(x) := u_n(x + k_n)$ so that

$$\int_{B(0, 1+\sqrt{N})} |v_n|^2 > \delta/2,$$

and, by $\mathbb{Z}^N$ invariance,

$$\varphi(v_n) \to c, \quad \varphi'(v_n) \to 0.$$

Going if necessary to a subsequence, we may assume that

$$v_n \rightharpoonup v \quad \text{in} \quad H^1(\mathbb{R}^N).$$

Since $v_n \to v$ in $L^2_{\text{loc}}(\mathbb{R}^N)$, $v \neq 0$. For every $w \in \mathcal{D}(\mathbb{R}^N)$, we have

$$\langle\varphi'(v), w\rangle = \lim_{n\to\infty} \langle\varphi'(v_n), w\rangle = 0.$$

Hence $\varphi'(v) = 0$ and v is a nontrivial solution of $(\mathcal{P})$. $\square$

Example 6.18. Assume that $V \in \mathcal{C}(\mathbb{R}^N)$ is 1-periodic in x_k, $1 \leq k \leq N$, and the linear operator

$$L : H^2(\mathbb{R}^N) \to L^2(\mathbb{R}^N) : u \mapsto -\Delta u + Vu$$

is invertible. Assume also that $2 < q < p < 2^*$. Then, for every $\lambda \geq 0$, problem

$$\begin{cases} -\Delta u + V(x)u = |u|^{p-2}u + \lambda|u|^{q-2}u, \\ u \in H^1(\mathbb{R}^N), \end{cases}$$

has a nontrivial solution. The existence of infinitely many solution was proved by Kryszewski and Szulkin.

Many papers are devoted to $(\mathcal{P})$ or to similar problems when F is strictly convex with respect to u (see [1], [28], [40], [43]). After a reduction it is possible to apply the mountain-pass theorem. In 1990, Hofer and Wysocki were able to construct homoclinics of a first order Hamiltonian system by using Fredholm operator theory and a linking argument. They do not assume the convexity of the Hamiltonian. Theorem 6.15 was first proved by Troestler and Willem under the supplementary assumption that, for some $2 < q < p < 2^*$, $|\partial_u f(x, u)| \leq c(|u|^{q-2} + |u|^{p-2})$.

Chapter 7

Generalized Kadomtsev-Petviashvili equation

7.1 Definition of solitary waves

This chapter is devoted to the existence of solitary waves of the generalized Kadomtsev-Petviashvili equation

(7.1) $$w_t + w_{xxx} + (f(w))_x = D_x^{-1} w_{yy},$$

where

$$D_x^{-1} h(x, y) := \int_{-\infty}^{x} h(s, y) ds.$$

A *solitary wave* is a solution of the form

$$w(t, x, y) = u(x - ct, y),$$

where $c > 0$ is fixed. Substituting in (7.1), we obtain

$$-cu_x + u_{xxx} + (f(u))_x = D_x^{-1} u_{yy}$$

or

$$(-u_{xx} + D_x^{-2} u_{yy} + cu - f(u))_x = 0.$$

7.2   Functional setting

In this section, $c > 0$ is fixed.

Definition 7.1. On $Y := \{g_x : g \in \mathcal{D}(\mathbb{R}^2)\}$ we define the inner product

$$(7.2) \qquad (u, v) := \int_{\mathbb{R}^2} [u_x v_x + D_x^{-1} u_y D_x^{-1} v_y + cuv]$$

and the corresponding norm

$$(7.3) \qquad \|u\| := \left(\int_{\mathbb{R}^2} [u_x^2 + (D_x^{-1} u_y)^2 + cu^2] \right)^{1/2}.$$

A function $u : \mathbb{R}^2 \to \mathbb{R}$ belongs to X if there exists $(u_n) \subset Y$ such that
(a) $u_n \to u$ a.e. on $\mathbb{R}^2$,
(b) $\|u_j - u_k\| \to 0,\ j, k \to \infty$.
The space X with inner product (7.2) and norm (7.3) is a Hilbert space.

Theorem 7.2. The following imbeddings are continuous:

$$X \subset L^p(\mathbb{R}^2), 2 \le p \le 6.$$

   Proof. See [17]. $\square$

Theorem 7.3. (de Bouard-Saut, 1994). The following imbeddings are compact:

$$X \subset L_{\text{loc}}^p(\mathbb{R}^2), 1 \le p < 6.$$

   Proof. Let (u_n) be a bounded sequence in X. There exists $(g_n) \subset L_{\text{loc}}^2$ such that $u_n = \partial_x g_n$ and $v_n := \partial_y g_n \in L^2$.

   Multiplying g_n by $\psi \in \mathcal{D}(\mathbb{R}^2)$ such that $0 \le \psi \le 1, \psi \equiv 1$ on B_R and supp $\psi \subset B_{2R}$, we may assume that supp $g_n \subset B_{2R}$. Going if necessary to a subsequence, we may assume that $u_n \rightharpoonup u = \partial_x g$ in X and replacing g_n by $g_n - g$, we may assume that $g = 0$. We denote by $\hat{u}(r, s)$ the Fourier transform of $u(x, y)$. We have, for $\varrho > 0$,

$$\int_{B_{2R}} |u_n|^2 = \int_{\mathbb{R}^2} |\hat{u}_n|^2$$
$$= \int_{\substack{|r| \le \varrho \\ |s| \le \varrho^2}} |\hat{u}_n|^2 + \int_{|r| > \varrho} |\hat{u}_n|^2 + \int_{\substack{|r| < \varrho \\ |s| > \varrho^2}} (\hat{u}_n)^2.$$

It is clear that

$$\int_{|r| > \varrho} |\hat{u}_n|^2 = \int_{|r| > \varrho} \frac{1}{4\pi^2 r^2} |\widehat{\partial_x u_n}|_2^2 \le \frac{1}{4\pi^2 \varrho^2} |\partial_x u_n|_2^2$$

and

$$\int_{\substack{|r| < \varrho \\ |s| > \varrho^2}} |\hat{u}_n|^2 = \int_{\substack{|r| < \varrho \\ |s| > \varrho^2}} \frac{r^2}{s^2} |\hat{v}_n|^2 \le \frac{1}{\varrho^2} |v_n|^2.$$

Let $\varepsilon > 0$. There exists $\rho > 0$ such that

$$\int_{|r|>\rho} |\hat{u}_n|^2 + \int_{\substack{|r|<\rho \\ |s|>\rho^2}} |\hat{u}_n|^2 \leq \varepsilon/2.$$

Since $u_n \rightharpoonup 0$ in $L^2(\mathbb{R}^2)$, we obtain

$$\hat{u}_n(r, s) = \int_{B_{2R}} u_n(x, y) c^{-2i\pi(xr+ys)} dx dy \to 0, n \to \infty$$

and

$$|\hat{u}_n(r, s)| \leq c_0 |u_n|_2 \leq c_1.$$

Lebesgue's dominated convergence theorem implies that

$$\int_{\substack{|r|\leq\rho \\ |s|\leq\rho^2}} |\hat{u}_n|^2 \to 0, n \to \infty.$$

Thus we have proved that $u_n \to 0$ in $L^2_{\text{loc}}(\mathbb{R}^2)$. By the preceding theorem, $u_n \to 0$ in $L^p_{\text{loc}}(\mathbb{R}^2), 1 \leq p < 6$. $\square$

Lemma 7.4. *If (u_n) is bounded in X and if*

$$\sup_{(x,y)\in\mathbb{R}^2} \int_{B(x,y;r)} |u_n|^2 \to 0, n \to \infty$$

then $u_n \to 0$ in $L^p(\mathbb{R}^2)$ for $2 < p < 6$.

Proof. Let $2 < s < 6$ and $u \in X$. The Hölder inequality and Theorem 7.2 imply that

$$
\begin{aligned}
|u|_{L^s(B(x,y;r))} &\leq |u|_{L^2(B(x,y;r))}^{1-\lambda} |u|_{L^6(B(x,y;r))}^{\lambda} \\
&\leq c_0 |u|_{L^2(B(x,y;r))}^{1-\lambda} \left(\int_{B(x,y;r)} [u_x^2 + (D_x^{-1}u_y)^2 + cu^2] \right)^{1/2}
\end{aligned}
$$

where $\lambda := \frac{s-2}{2}\frac{3}{s}$. Choosing $\lambda = 2/s$, we obtain

$$\int_{B(x,y;r)} |u|^s \leq c_0^s |u|_{L^2(B(x,y;r))}^{(1-\lambda)s} \int_{B(x,y;r)} [u_x^2 + (D_x^{-1}u_y)^2 + cu^2].$$

Now, covering $\mathbb{R}^2$ by balls of radius r in such a way that each point of $\mathbb{R}^2$ is contained in at most 3 balls, we find

$$\int_{\mathbb{R}^2} |u|^s \leq 3c_0^s \sup_{(x,y)\in\mathbb{R}^2} \left(\int_{B(x,y;r)} |u|^2 \right)^{(1-\lambda)s/2} \int_{\mathbb{R}^2} [y_x^2 + (D_x^{-1}u_y)^2 + cu^2].$$

Under the assumptions of the lemma, $u_n \to 0$ in $L^s(\mathbb{R}^2)$. Since $2 < s < 6$, $u_n \to 0$ in $L^p(\mathbb{R}^2)$ for $2 < p < 6$, by the Hölder inequality and Theorem 7.2. $\square$

7.3　Existence of solitary waves

We consider the problem

$$(\mathcal{P}) \qquad \begin{cases} (-u_{xx} + D_x^{-2}u_{yy} + cu - f(u))_x = 0, \\ u \in X, \end{cases}$$

where $c > 0$ is fixed. We define

$$F(u) := \int_0^u f(s)ds$$

and we assume that

(f_1) $f \in C^1(\mathbb{R}, \mathbb{R})$, $f(0) = 0$ and for some $3 < p < 6, c_0 > 0$,

$$|f'(u)| \le c_0(|u| + |u|^{p-2}),$$

(f_2) there exists $v \in X$ such that

$$F(\lambda v)/\lambda^2 \to +\infty, \lambda \to +\infty,$$

(f_3) there exists $\alpha > 2$ such that, for $u \in \mathbb{R}$,

$$\alpha F(u) \le uf(u).$$

The weak solutions of $(\mathcal{P})$ are the critical points of the function φ defined on X by

$$\varphi(u) := \int_{\mathbb{R}^2} \left[\frac{1}{2}(u_x^2 + (D_x^{-1}u_y)^2 + cu^2) - F(u) \right].$$

By assumption (f_1), $\varphi \in C^2(X, \mathbb{R})$.

Lemma 7.5. *Under assumptions (f_1) and (f_2) there exists $e \in X$ and $r > 0$ such that $\|e\| \ge r$ and*

$$b := \inf_{\|u\|=r} \varphi(u) > \varphi(0) \ge \varphi(e).$$

Proof. Assumption (f_1) implies the existence of $c_1 > 0$ such that, on $\mathbb{R}$,

$$F(u) \le \frac{|u|^2}{4} + c_1|u|^p.$$

Hence we obtain

$$\begin{aligned} \varphi(u) &\ge \frac{\|u\|^2}{2} - \int \left(\frac{|u|^2}{4} + c_1|u|^p \right) \\ &\ge \frac{\|u\|^2}{4} - c_1|u|_p^p. \end{aligned}$$

By Theorem 7.2, there exists $r > 0$ such that

$$b := \inf_{\|u\|=r} \varphi > 0.$$

It follows from assumption (f_2) that

$$\varphi(\lambda v) \to -\infty, \lambda \to +\infty.$$

Hence there exists $\lambda > 0$ such that

$$\|\lambda v\| > r, 0 = \varphi(c) \geq \varphi(\lambda v). \qquad \square$$

We define

$$d := \inf_{\gamma \in \Gamma} \max_{t \in [0,1]} \varphi(\gamma(t)),$$

$$\Gamma := \{\gamma \in C([0,1], X) : \gamma(0) = 0, \gamma(1) = e\}.$$

Lemma 7.6. *Under assumptions (f_1)-(f_2), $d \geq b$ and there exists a sequence $(u_n) \subset X$ such that*

$$\varphi(u_n) \to d, \varphi'(u_n) \to 0.$$

Proof. Let $\gamma \in \Gamma$. Since $\|\gamma(0)\| = 0$, $\|\gamma(1)\| > r$, there exists $t \in [0,1]$ such that $\|\gamma(t)\| = r$ and so

$$d \geq b > \varphi(0) \geq \varphi(e).$$

It suffices then to apply Theorem 2.8. $\square$

Theorem 7.7. *Under assumptions (f_1)-(f_3), problem $(\mathcal{P})$ has a nontrivial solution.*

Proof. 1) Let (u_n) be the sequence given by the preceding lemma. For n big enough, we have, by (f_3),

$$
\begin{aligned}
d + 1 + \|u_n\| &\geq \varphi(u_n) - \alpha^{-1}\langle \varphi'(u_n), u_n \rangle \\
&= \left(\frac{1}{2} - \frac{1}{\alpha}\right)\|u_n\|^2 + \int [\alpha^{-1}u_n f(u_n) - F(u_n)] \\
&\geq \left(\frac{1}{2} - \frac{1}{\alpha}\right)\|u_n\|^2.
\end{aligned}
$$

Thus (u_n) is bounded in X.

2) If

$$\delta := \varlimsup_{n \to \infty} \sup_{(x,y) \in \mathbb{R}^2} \int_{B(x,y;1)} |u_n|^2 = 0,$$

then by Lemma 7.4, $u_n \to 0$ in $L^s(\mathbb{R}^2)$, for $2 < s < 6$. It follows that

$$0 < d = \varphi(u_n) - \frac{1}{2}\langle \varphi'(u_n), u_n \rangle + o(1)$$
$$= \int \left[\frac{1}{2} u_n f(u_n) - F(u_n) \right] + o(1) = o(1).$$

This is a contradiction. Thus $\delta > 0$.

3) Going if necessary to a subsequence, we can assume the existence of $((x_n, y_n)) \subset \mathbb{R}^2$ such that

$$\int_{B(x_n, y_n; 1)} |u_n|^2 > \delta/2.$$

Let us define $v_n(x, y) := u_n(x + x_n, y + y_n)$ so that

$$\int_{B(0;1)} |v_n|^2 > \delta/2.$$

Going if necessary to a subsequence, we may assume that

$$v_n \rightharpoonup v \text{ in } X.$$

By Theorem 7.3, $v_n \to v$ in $L^2_{\text{loc}}(\mathbb{R}^2)$ and so $v \neq 0$. For every $w \in Y$, we have

$$\langle \varphi'(v), w \rangle = \lim_{n \to \infty} \langle \varphi'(v_n), w \rangle = 0.$$

Hence $\varphi'(v) = 0$ and v is a nontrivial solution of $(\mathcal{P})$. $\square$

Example 7.8. The following nonlinearities satisfy assumptions (f_1)-(f_3):

$$f(u) := \lambda u^4 - u^3, \lambda \in \mathbb{R} \backslash \{0\},$$
$$f(u) := u^3 + \lambda u^2, \lambda \in \mathbb{R}.$$

The case $f(u) = u^{p-1}$, $p = 3, 4, 5$, was solved by de Bouard and Saut using constrained minimization.

7.4   Variational identity

In this section, we derive some non existence results from a variational identity. We give first a formal argument. We define on X the transformations

$$T(t)u(x, y) := u(x/t, y/t^2), t > 0$$

and the functional

$$\varphi(u) := \int_{\mathbb{R}^2} \left[\frac{1}{2}(u_x^2 + (D_x^{-1} u_y)^2) - G(u) \right].$$

We have that

$$\varphi(T(t)u) = \frac{t}{2} \int_{\mathbb{R}^2} (u_x^2 + (D_x^{-1} u_y)^2) - t^3 \int_{\mathbb{R}^2} G(u)$$

and

$$\frac{\partial}{\partial t}\Big|_{t=1} \varphi(T(t)u) = \frac{1}{2} \int_{\mathbb{R}^2} (u_x^2 + (D_x^{-1} u_y)^2) - 3 \int_{\mathbb{R}^2} G(u).$$

If u is a critical point of φ, we conjecture that

$$\frac{1}{2} \int_{\mathbb{R}^2} (u_x^2 + (D_x^{-1} u_y)^2) = 3 \int_{\mathbb{R}^2} G(u).$$

Let $g \in C^1(\mathbb{R}, \mathbb{R})$ be such that $g(0) = 0$ and define

$$G(u) := \int_0^u g(s) ds.$$

Theorem 7.9. *Any solution of*

$$\begin{cases} (-u_{xx} + D_x^{-2} u_{yy} - g(u))_x = 0, \\ u \in X \cap H_{loc}^2(\mathbb{R}^2), \\ G(u), g(u)u, g(u) D_x^{-1} u_y \in L^1(\mathbb{R}^2), \end{cases}$$

satisfies

(7.4) $$\frac{1}{2} \int_{\mathbb{R}^2} (u_x^2 + (D_x^{-1} u_y)^2) = 3 \int_{\mathbb{R}^2} G(u).$$

Proof. Let $\psi \in \mathcal{D}(\mathbb{R})$ be such that $0 \leq \psi \leq 1$, $\psi(r) = 1$ for $r \leq 1$ and $\psi(r) = 0$ for $r \geq 2$. Define, on $\mathbb{R}^2$,

$$\psi_n(x, y) := \psi\left(\frac{x^2 + y^2}{n^2}\right).$$

We have, for every n,

(7.5) $$\int (-u_{xx} + D_x^{-2} u_{yy} - g(u))(\psi_n x u)_x = 0.$$

Integrating by parts, we find

$$-\int u_{xx}(\psi_n x u)_x = -\int u_{xx}(\psi_{n,x} x u + \psi_n u + \psi_n x u_x)$$

$$= \int \left(\frac{3}{2} u_x^2 [\psi_{n,x} x + \psi_n] + 2\psi_{n,x} u u_x + \psi_{n,xx} x u u_x\right).$$

Lebesgue dominated convergence theorem implies that

(7.6) $$-\int u_{xx}(\psi_n x u)_x = \frac{3}{2} \int u_x^2 + o(1).$$

Similarly we obtain

(7.7)
$$\int D_x^{-2} u_{yy} (\psi_n x u)_x = -\frac{1}{2} \int (D_x^{-1} u_y)^2 + o(1)$$

and

(7.8)
$$-\int g(u)(\psi_n x u)_x = \int [G(u) - g(u)u] + o(1).$$

We infer from (7.5)-(7.8) that

$$\int \left[\frac{3}{2} u_x^2 - \frac{1}{2} (D_x^{-1} u_y)^2 + G(u) - g(u)u \right] = 0.$$

Since

$$\int [u_x^2 + (D_x^{-1} u_y)^2] = \int g(u)u,$$

we obtain

(7.9)
$$\int [\frac{1}{2} u_x^2 - \frac{3}{2} (D_x^{-1} u_y)^2 + G(u)] = 0.$$

For every n, we have also

$$\int (-u_{xx} + D_x^{-2} u_{yy} - g(u))(\psi_n y D_x^{-1} u_y)_x = 0.$$

Integrating by parts and using the Lebesgue dominated convergence theorem, we obtain

(7.10)
$$\int \left[-\frac{1}{2} u_x^2 + \frac{1}{2} (D_x^{-1} u_y)^2 + G(u) \right] = 0.$$

Formula (7.4) follows from (7.9) and (7.10). $\square$

Corollary 7.10. Let $c > 0$ and $p := p_1/p_2$, where p_1 and p_2 are relatively prime and p_2 is odd. If $p \geq 6$, then 0 is the only solution of

$$\begin{cases} (-u_{xx} + D_x^{-2} u_{yy} + cu - u^{p-1})_x = 0, \\ u \in X \cap H_{loc}^2(\mathbb{R}^2) \cap L^p(\mathbb{R}^2), \\ u^{p-1} D_x^{-1} u_y \in L^1(\mathbb{R}^2). \end{cases}$$

   Proof. Formula (7.4) leads to

$$0 = \int \left[\frac{1}{2}(u^p - cu^2) - 3\left(\frac{u^p}{p} - \frac{cu^2}{2} \right) \right]$$

$$= \int \left[\left(\frac{1}{2} - \frac{3}{p} \right) u^p + cu^2 \right].$$

Since $\int u^p = \|u\|^2$, it follows that $u = 0$ if $p \geq 6$. $\square$

   When $p \geq 6$ is an integer, the above result was proved by de Bouard and Saut.

Chapter 8

Representation of Palais-Smale sequences

8.1 Invariance by translations

In this chapter, we describe losses of compactness in some variational problems. Minimizing sequences were considered by Pierre-Louis Lions in [50] and [51]. Palais-Smale sequences were studied by many authors (see the bibliography of [21]). Of course the Ekeland principle allows a reduction of minimizing sequences to Palais-Smale sequences.

We consider the "limit functional" ψ associated to some functional φ. The method is to iterate the following procedure:

a) (u_n^1) is a Palais-Smale sequence for ψ which converges weakly to 0,

b) we replace (u_n^1) by a sequence $v_n^1(x) := u_n^1(x + y_n^1)$ which converges weakly to $v_1 \neq 0$,

c) $u_n^2(x) := u_n^1(x) - v_1(x - y_n^1)$ is a Palais-Smale sequence for ψ which converges weakly to 0.

In this section, we assume that

(A) Ω is a smooth domain of $\mathbb{R}^N$ with a bounded complement, $2 < p < 2^*$, $\lambda > 0$, $a \in \mathcal{C}(\Omega)$, $\inf_\Omega a > 0$ and $\lim_{|x| \to \infty} a(x) = 1$.

We define the functionals

$$\varphi(u) := \int_\Omega \left[\frac{|\nabla u|^2}{2} + a\frac{u^2}{2} - \lambda\frac{|u|^p}{p} \right], u \in H_0^1(\Omega),$$

$$\psi(u) := \int_{\mathbb{R}^N} \left[\frac{|\nabla u|^2}{2} + \frac{u^2}{2} - \lambda\frac{|u|^p}{p} \right], u \in H^1(\mathbb{R}^N),$$

we denote by $\|.\|$ the usual norm on $H^1(\mathbb{R}^N)$.

Lemma 8.1. *If $2 < p < 2^*$ and if $u_n \rightharpoonup u$ in $H^1(\mathbb{R}^N)$ then*

$$|u_n|^{p-2}u_n - |u_n - u|^{p-2}(u_n - u) \to |u|^{p-2}u \text{ in } H^{-1}(\mathbb{R}^N).$$

Proof. Let us define $f(v) := |v|^{p-2}v$. By the mean value theorem, we have, a.e. on $\mathbb{R}^N$,

$$|f(u_n) - f(u_n - u)| \le (p-1)[|u_n| + |u|]^{p-2}|u|.$$

For $R > 0$ and $w \in \mathcal{D}(\mathbb{R}^N)$, we obtain from the Hölder inequality,

$$\left|\int_{|x|>R}|f(u_n) - f(u_n - u)|w\right| \le c_1 \left[|u_n|_p^{p-2} + |u|_p^{p-2}\right]|w|_p \left[\int_{|x|>R}|u|^p\right]^{\frac{1}{p}}$$

$$\le c_2\|w\|\left[\int_{|x|>R}|u|^p\right]^{\frac{1}{p}}.$$

We have also that

$$\left|\int_{|x|>R}f(u)w\right| \le |w|_p\left[\int_{|x|>R}|u|^p\right]^{\frac{p-1}{p}} \le c_3\|w\|\left[\int_{|x|>R}|u|^p\right]^{\frac{p-1}{p}}.$$

Thus, for every $\varepsilon > 0$, there exists $R > 0$ such that, for every $w \in \mathcal{D}(\mathbb{R}^N)$,

$$\left|\int_{|x|>R}(f(u_n) - f(u_n - u) - f(u))\,w\right| \le \varepsilon\|w\|.$$

It follows from the Rellich theorem and Theorem A.4 that

$$f(u_n) - f(u_n - u) \to f(u) \text{ in } L^r(B(0, R))$$

where $r := (p-1)/p$. Since

$$\left|\int_{|x|<R}(f(u_n) - f(u_n - u) - f(u))w\right|$$
$$\le |w|_p|f(u_n) - f(u_n - u) - f(u)|_{L^r(B(0,R))}$$
$$\le c_3\|w\|\,|f(u_n) - f(u_n - u) - f(u)|_{L^r(B(0,R))}$$

the proof is complete. $\square$

Lemma 8.2. *If*

$$u_n \rightharpoonup u \text{ in } H_0^1(\Omega),$$
$$u_n \to u \text{ a.e. on } \Omega,$$
$$\varphi(u_n) \to c,$$
$$\varphi'(u_n) \to 0 \text{ in } H^{-1}(\Omega),$$

then $\varphi'(u) = 0$ and $v_n := u_n - u$ is such that

$$||v_n||^2 = ||u_n||^2 - ||u||^2 + o(1),$$
$$\psi(v_n) \to c - \varphi(u),$$
$$\psi'(v_n) \to 0 \text{ in } H^{-1}(\Omega).$$

Proof. 1) It is clear that

$$||v_n||^2 = ||u_n||^2 - ||u||^2 + o(1).$$

2) Since $\lim_{|x| \to \infty} a(x) = 1$ and $v_n \to 0$ in $L_{loc}^2(\Omega)$, we have

$$\int_\Omega a(x)v_n^2 = \int_\Omega v_n^2 + o(1).$$

According to the Brézis-Lieb lemma, we obtain

$$\begin{aligned}
\psi(v_n) &= \varphi(v_n) + o(1) \\
&= \varphi(u_n) - \varphi(u) + o(1) \\
&= c - \varphi(u) + o(1).
\end{aligned}$$

3) Since $u_n \rightharpoonup u$ in $H_0^1(\Omega)$ and $\varphi'(u_n) \to 0$ in $H^{-1}(\Omega)$, it is easy to verify that $\varphi'(u) = 0$. The preceding lemma implies that

$$\begin{aligned}
\psi'(v_n) &= \varphi'(v_n) + o(1) \\
&= \varphi'(u_n) - \varphi'(u) + o(1) \\
&= o(1). \quad \square
\end{aligned}$$

Lemma 8.3. *If $|y_n| \to \infty$ and*

$$u_n(. + y_n) \rightharpoonup u \text{ in } H^{-1}(\mathbb{R}^N),$$
$$u_n(. + y_n) \to u \text{ a.e. on } \mathbb{R}^N,$$
$$\psi(u_n) \to c,$$
$$\psi'(u_n) \to 0 \text{ in } H^{-1}(\Omega),$$

then $\psi'(u) = 0$ and $v_n := u_n - u(. - y_n)$ is such that

$$||v_n||^2 = ||u_n||^2 - ||u||^2 + o(1),$$
$$\psi(v_n) \to c - \psi(u),$$
$$\psi'(v_n) \to 0 \text{ in } H^{-1}(\Omega).$$

Proof. 1) It is clear that

$$\begin{aligned} ||v_n||^2 &= ||u_n(. + y_n) - u||^2 \\ &= ||u_n||^2 - ||u||^2 + o(1). \end{aligned}$$

2) According to the Brézis-Lieb lemma, we obtain

$$\begin{aligned} \psi(v_n) &= \psi(u_n(. + y_n) - u) \\ &= \psi(u_n(. + y_n)) - \psi(u) + o(1) \\ &= \psi(u_n) - \psi(u) + o(1) \\ &= c - \psi(u) + o(1). \end{aligned}$$

3) Since $u_n(. + y_n) \to u$ in $H^1(\mathbb{R}^N)$, $\psi'(u_n) \to 0$ in $H^{-1}(\Omega)$ and $(y_n) \to \infty$, it is easy to verify that $\psi'(u) = 0$. Lemma 8.1 implies that

$$\begin{aligned} \psi'(v_n) &= \psi'(u_n(. + y_n) - u) \\ &= \psi'(u_n(. + y_n)) - \psi'(u) + o(1) \\ &= \psi'(u_n) + o(1) = o(1). \quad \square \end{aligned}$$

Theorem 8.4. (Benci-Cerami, 1987). *Under assumption* (A), *let* $(u_n) \subset H_0^1(\Omega)$ *be such that*

$$\varphi(u_n) \to c, \quad \varphi'(u_n) \to 0 \text{ in } H^{-1}(\Omega).$$

Then, replacing (u_n) *if necessary by a subsequence, there exists a solution* $v_0 \in H_0^1(\Omega)$ *of*

$$-\Delta u + a\, u = \lambda |u|^{p-2}u,$$

a finite sequence $(v_1, ..., v_k) \subset H^1(\mathbb{R}^N)$ *of solutions of*

$$-\Delta u + u = \lambda |u|^{p-2}u,$$

and k *sequences* (y_n^j) *satisfying*

$$|y_n^j| \to \infty, |y_n^j - y_n^{j'}| \to \infty, j \neq j', n \to \infty,$$
$$||u_n - v_0 - \sum_{j=1}^k v_j(. - y_n^j)|| \to 0,$$
$$||u_n||^2 \to \sum_{j=0}^k ||v_j||^2,$$
$$\varphi(v_0) + \sum_{j=1}^k \psi(v_j) = c.$$

Moreover, if $k \geq 1$ and $u_n \geq 0$ a.e. on Ω for all n, then $v_j > 0$ on $\mathbb{R}^N$ for $1 \leq j \leq k$.

Proof. 1) As in the proof of Lemma 1.20, (u_n) is bounded. We may assume that $u_n \rightharpoonup v_0$ in $H_0^1(\Omega)$ and $u_n \to v_0$ a.e. on Ω. By Lemma 8.2, $\varphi'(v_0) = 0$ and $u_n^1 := u_n - v_0$ is such that

(1)
$$\|u_n^1\|^2 = \|u_n\|^2 - \|v_0\|^2 + o(1),$$
$$\psi(u_n^1) \to c - \varphi(v_0),$$
$$\psi'(u_n^1) \to 0 \text{ in } H^{-1}(\Omega).$$

2) Let us define

$$\delta := \varlimsup_{n \to \infty} \sup_{y \in \mathbb{R}^N} \int_{B(y,1)} |u_n^1|^2.$$

If $\delta = 0$, Lemma 1.21 implies that $u_n^1 \to 0$ in $L^p(\mathbb{R}^N)$. Since $\psi'(u_n^1) \to 0$, it follows that $u_n^1 \to 0$ in $H^1(\mathbb{R}^N)$ and the proof is complete. If $\delta > 0$, we may assume the existence of $(y_n^1) \subset \mathbb{R}^N$ such that

$$\int_{B(y_n^1,1)} |u_n^1|^2 > \delta/2.$$

Let us define $v_n^1 := u_n^1(\,.\, + y_n^1)$. We may assume that $v_n^1 \rightharpoonup v_1$ in $H^1(\mathbb{R}^N)$ and $v_n^1 \to v_1$ a.e. on $\mathbb{R}^N$. Since

$$\int_{B(0,1)} |v_n^1|^2 > \delta/2$$

it follows from the Rellich theorem that

$$\int_{B(0,1)} |v^1|^2 \geq \delta/2$$

and $v_1 \neq 0$. But $u_n^1 \rightharpoonup 0$ in $H^1(\mathbb{R}^N)$, so that (y_n^1) is unbounded. We may assume that $|y_n^1| \to \infty$. Finally, by (1) and the preceding lemma, $\psi'(v_1) = 0$ and $u_n^2 := u_n^1 - v^1(\,.\, - y_n^1)$ satisfies

(2)
$$\|u_n^2\|^2 = \|u_n\|^2 - \|v_0\|^2 - \|v_1\|^2 + o(1),$$
$$\psi(u_n^2) \to c - \varphi(v_0) - \psi(v_1),$$
$$\psi'(u_n^2) \to 0 \text{ in } H^{-1}(\Omega).$$

3) Any nontrivial critical point of ψ satisfies

$$S_p |u|_p^2 \leq \|u\|^2 = \lambda |u|_p^p$$

so that

(3)
$$\psi(u) \geq c^* := \lambda \left(\frac{1}{2} - \frac{1}{p} \right) \left(\frac{S_p}{\lambda} \right)^{\frac{p}{p-2}}.$$

Iterating the above procedure we construct sequences (v_j) and (y_j^n). Since, for every j, $\psi(v_j) \geq c^*$, the iteration must terminate at some finite index k.

4) If $u_n \geq 0$ a.e. on Ω for all n, it suffices to use the functionals

$$\varphi_+(u) := \int_\Omega \left[\frac{|\nabla u|^2}{2} + a\frac{u^2}{2} - \lambda\frac{u_+^p}{p} \right], u \in H_0^1(\Omega),$$

$$\psi_+(u) := \int_{\mathbb{R}^N} \left[\frac{|\nabla u|^2}{2} + \frac{u^2}{2} - \lambda\frac{u_+^p}{p} \right], u \in H^1(\mathbb{R}^N),$$

and the maximum principle. $\square$

We need the following version of the Ekeland variational principle.

Theorem 8.5. *Let X be a Banach space and let $G \in C^2(X, \mathbb{R})$ be such that, for every $v \in V := \{v \in X : G(v) = 1\}$, $G'(v) \neq 0$. Let $F \in C^1(X, \mathbb{R})$ be bounded below on V, $v \in V$ and $\varepsilon, \delta > 0$. If*

$$F(v) \leq \inf_V F + \varepsilon$$

there exists $u \in V$ such that

$$F(u) \leq \inf_V F + 2\varepsilon, \min_{\lambda \in \mathbb{R}} ||F'(u) - \lambda G'(u)|| \leq 8\varepsilon/\delta, ||u - v|| \leq 2\delta.$$

Proof. It suffices to apply Lemma 5.15 with $S := \{-v\}$ and $c := \inf_V \varphi$. $\square$

We consider the following minimization problems where $2 < p < 2^*$

$$S_a := \inf_{\substack{u \in H_0^1(\Omega) \\ |u|_p = 1}} \int_\Omega [|\nabla u|^2 + a\,u^2],$$

$$S_p := \inf_{\substack{u \in H^1(\mathbb{R}^N) \\ |u|_p = 1}} \int_{\mathbb{R}^N} [|\nabla u|^2 + u^2].$$

Theorem 8.6. (P.L. Lions, 1984). *Under assumption (A), if $S_a < S_p$, then every sequence $(u_n) \subset H_0^1(\Omega)$ satisfying*

$$|u_n|_p = 1, \int_\Omega [|\nabla u_n|^2 + a\,u_n^2] \to S_a, n \to \infty,$$

contains a convergent subsequence. In particular, there exists a minimizer for S_a.

Proof. Applying the Ekeland principle to

$$G(u) := \int_\Omega \frac{|u|^p}{p}, \; F(u) := \int_\Omega \left[\frac{|\nabla u|^2}{2} + a\frac{u^2}{2} \right]$$

we may assume that

$$\min_{\lambda \in \mathbb{R}} ||F'(u_n) - \lambda G'(u_n)|| \to 0, n \to \infty.$$

It is then easy to verify that, for $\lambda := S_a$,

$$F'(u_n) - \lambda G'(u_n) \to 0, F(u_n) - \lambda G(u_n) \to \left(\frac{1}{2} - \frac{1}{p} \right) \lambda.$$

Hence (u_n) is a Palais-Smale sequence of

$$\varphi(u) := \int_\Omega \left[\frac{|\nabla u|^2}{2} + a\frac{u^2}{2} - \lambda\frac{|u|^p}{p} \right].$$

By (3) any nontrivial critical point u of

$$\psi(u) := \int_{\mathbb{R}^N} \left[\frac{|\nabla u|^2}{2} + \frac{u^2}{2} - \lambda\frac{|u|^p}{p} \right]$$

satisfies

$$\psi(u) \geq c^* := \lambda \left(\frac{1}{2} - \frac{1}{p} \right) \left(\frac{S_p}{\lambda} \right)^{\frac{p}{p-2}}.$$

If $\lambda = S_a < S_p$, then $c^* > \lambda \left(\frac{1}{2} - \frac{1}{p} \right)$ and Theorem 8.4 implies that $u_n \to v_0$ in $H_0^1(\Omega)$. $\square$

We now give a short proof of Theorem 1.34.

Theorem 8.7. (P.L. Lions, 1984). *Let $(u_n) \subset H^1(\mathbb{R}^N)$ be such that*

$$|u_n|_p = 1, \int_{\mathbb{R}^N} [|\nabla u_n|^2 + u_n^2] \to S_p, n \to \infty.$$

Then there exists a sequence $(y_n) \subset \mathbb{R}^N$ such that $u_n(. + y_n)$ contains a convergent subsequence. In particular, there exists a minimizer for S_p.

Proof. Using the Ekeland principle, we may assume that

$$\psi'(u_n) \to 0, \psi(u_n) \to \left(\frac{1}{2} - \frac{1}{p} \right) \lambda$$

where $\lambda := S_p$ and

$$\psi(u) := \int_{\mathbb{R}^N} \left[\frac{|\nabla u|^2}{2} + \frac{u^2}{2} - \frac{|u|^p}{p} \right].$$

By (3), any nontrivial critical point u of ψ satisfies

$$\psi(u) \geq c^* := \left(\frac{1}{2} - \frac{1}{p}\right)\lambda.$$

Theorem 8.4 implies that either $||u_n - v_0|| \to 0$ or $||u_n - v_1(. - y_n^1)||$ $\to 0$. $\square$

8.2   Symmetric domains

When Ω and a are invariant by a group of orthogonal transformations, it is possible to improve Theorem 8.6.

Let G be a subgroup of $\mathbf{O}(N)$. Let Ω be an invariant domain of $\mathbb{R}^N$ and let $a \in C(\Omega)$ be an invariant function satisfying (A). As in Definition 1.23, the action of G on $H_0^1(\Omega)$ is defined by

$$gu(x) := u(g^{-1}x).$$

The subspace of invariant functions is defined by

$$H_{0,G}^1(\Omega) := \{u \in H_0^1(\Omega) : gu = u, \forall g \in G\}.$$

We define also

$$m(G) := \inf_{|x|=1} \#\{gx : g \in G\},$$

$$S_G := \inf_{\substack{u \in H_{0,G}^1(\Omega) \\ |u|_p=1}} \int_\Omega [|\nabla u|^2 + a\, u^2].$$

Theorem 8.8. (P.L. Lions, 1985). *Under assumption (A), if Ω and a are G-invariant and if $S_G < m(G)^{\frac{p-2}{p}} S_p$, then every sequence $(u_n) \subset H_{0,G}^1(\Omega)$ satisfying*

$$|u_n|_p = 1, \int_\Omega [|\nabla u_n|^2 + a\, u_n^2] \to S_G, n \to \infty,$$

contains a convergent subsequence. In particular there exists a minimizer for S_G.

   Proof.   Using the Ekeland principle and the principle of symmetric criticality, we may assume that

$$\varphi'(u_n) \to 0, \quad \varphi(u_n) \to \lambda\left(\frac{1}{2} - \frac{1}{p}\right),$$

where $\lambda := S_G$ and

$$\varphi(u) := \int_\Omega \left[\frac{|\nabla u|^2}{2} + a\frac{u^2}{2} - \lambda\frac{|u|^p}{p}\right].$$

By (3), any nontrivial critical point u of

$$\psi(u) := \int_{\mathbb{R}^N} \left[\frac{|\nabla u|^2}{2} + \frac{u^2}{2} - \lambda \frac{|u|^p}{p} \right]$$

satisfies

$$\psi(u) \geq c^* := \lambda \left(\frac{1}{2} - \frac{1}{p} \right) \left(\frac{S_p}{\lambda} \right)^{\frac{p}{p-2}}.$$

We conclude from Theorem 8.4 and from the invariance of the problem that either $k = 0$ or $k \geq m(G)$. If $k \geq m(G)$ then

$$c^* m(G) \leq \lambda \left(\frac{1}{2} - \frac{1}{p} \right).$$

This is not possible since, by assumption,

$$c^* m(G) > \lambda \left(\frac{1}{2} - \frac{1}{p} \right).$$

Thus $k = 0$ and $u_n \to v_0$ in $H_0^1(\Omega)$. $\square$

8.3 Invariance by dilations

In this section, we assume that

(B) Ω is a smooth bounded domain of $\mathbb{R}^N$, $N \geq 3$, $\lambda > 0$ and $a \in \mathcal{C}^\infty(\Omega) \cap L^{N/2}(\Omega)$ is such that

$$\inf_{\substack{u \in \mathcal{D}_0^{1,2}(\Omega) \\ |\nabla u|_2 = 1}} \int_\Omega (|\nabla u|^2 + a\, u^2) > 0.$$

We define the functionals

$$\varphi(u) := \int_\Omega \left[\frac{|\nabla u|^2}{2} + a \frac{u^2}{2} - \lambda \frac{|u|^{2^*}}{2^*} \right], u \in \mathcal{D}_0^{1,2}(\Omega),$$

$$\psi(u) := \int_{\mathbb{R}^N} \left[\frac{|\nabla u|^2}{2} - \lambda \frac{|u|^{2^*}}{2^*} \right], u \in \mathcal{D}^{1,2}(\mathbb{R}^N).$$

On $\mathcal{D}^{1,2}(\mathbb{R}^N)$, we define the norm $\|u\| := |\nabla u|_2$.

Lemma 8.9. *If $u_n \rightharpoonup u$ in $\mathcal{D}^{1,2}(\mathbb{R}^N)$ and $u \in L_{\text{loc}}^\infty(\mathbb{R}^N)$ then*

$$|u_n|^{2^*-2} u_n - |u_n - u|^{2^*-2}(u_n - u) \to |u|^{2^*-2} u \text{ in } (\mathcal{D}^{1,2}(\mathbb{R}^N))'.$$

Proof. Let us define $f(v) := |v|^{2^*-2} v$.

By the mean value theorem, we have, a.e. on $\mathbb{R}^N$,

$$|f(u_n) - f(u_n - u)| \leq (2^* - 1)[|u_n| + |u|]^{2^*-2}|u|.$$

As in the proof of Lemma 8.1, for every $\varepsilon > 0$, there exists $R > 0$ such that, for every $w \in \mathcal{D}(\mathbb{R}^N)$,

$$\left| \int_{|x|>R} (f(u_n) - f(u_n - u) - f(u))w \right| \leq \varepsilon \|w\|.$$

Let us define $M := \sup_{B(0,R)} |u|$, so that a.e. on $B(0, R)$,

$$|f(u_n) - f(u_n - u)| \leq (2^* - 1)[|u_n| + M]^{2^*-2}M.$$

It follows from the Rellich theorem and Theorem A.4 that

$$f(u_n) - f(u_n - u) \to f(u) \text{ in } L^r(B(0, R))$$

where $r := 2N/5$. Since

$$\left| \int_{|x|>R} (f(u_n) - f(u_n - u) - f(u))w \right|$$
$$\leq |w|_{L^s(B(0,R))}|f(u_n) - f(u_n - u) - f(u)|_{L^r(B(0,R))}$$
$$\leq c_1\|w\| \, |f(u_n) - f(u_n - u) - f(u)|_{L^r(B(0,R))}$$

where $s := 2N/(2N - 5)$, the proof is complete. $\square$

Lemma 8.10. *If*

$$u_n \rightharpoonup u \text{ in } \mathcal{D}_0^{1,2}(\Omega),$$
$$u_n \to u \text{ a.e. on } \Omega,$$
$$\varphi(u_n) \to c,$$
$$\varphi'(u_n) \to 0 \text{ in } (\mathcal{D}_0^{1,2}(\Omega))',$$

then $\varphi'(u) = 0$ and $v_n := u_n - u$ is such that

$$\|v_n\|^2 = \|u_n\|^2 - \|u\|^2 + o(1),$$
$$\psi(v_n) \to c - \varphi(u),$$
$$\psi'(v_n) \to 0 \text{ in } (\mathcal{D}_0^{1,2}(\Omega))'.$$

Proof. 1) Since $v_n \rightharpoonup 0$ in $\mathcal{D}_0^{1,2}(\Omega)$, it is clear that

$$\|v_n\|^2 = \|u_n\|^2 - \|u\|^2 + o(1).$$

2) Since (v_n) is bounded in $L^{2^*}(\Omega)$, (v_n^2) is bounded in $L^{N/(N-2)}(\Omega)$ and so (see [90])

$$v_n^2 \rightharpoonup 0 \text{ in } L^{N/(N-2)}.$$

It follows that

$$\int_\Omega a(x) v_n^2 = o(1).$$

According to the Brézis-Lieb lemma, we obtain

$$\begin{aligned} \psi(v_n) &= \varphi(v_n) + o(1) \\ &= \varphi(u_n) - \varphi(u) + o(1) \\ &= c - \varphi(u) + o(1). \end{aligned}$$

3) Since $u_n \rightharpoonup u$ in $\mathcal{D}_0^{1,2}(\Omega)$ and $\varphi'(u_n) \to 0$, it is easy to verify that $\varphi'(u) = 0$. By the argument of Lemma 1.30, $u \in C^2(\mathbb{R}^N)$. The preceding lemma implies that

$$\begin{aligned} \psi'(v_n) &= \varphi'(v_n) + o(1) \\ &= \varphi'(u_n) - \varphi'(u) + o(1) \\ &= o(1). \quad \square \end{aligned}$$

Lemma 8.11. *Let* $(y_n) \subset \Omega$ *and* $(\lambda_n) \subset]0, \infty[$ *be such that*

(4)
$$\frac{1}{\lambda_n} \operatorname{dist}(y_n, \partial\Omega) \to \infty.$$

If the sequence (u_n) *and the rescaled sequence*

$$v_n(x) := \lambda_n^{(N-2)/2} u_n(\lambda_n x + y_n)$$

are such that

$$\begin{aligned} &v_n \rightharpoonup v \text{ in } \mathcal{D}^{1,2}(\mathbb{R}^N), \\ &v_n \to v \text{ a.e. on } \mathbb{R}^N, \\ &\psi(u_n) \to c, \\ &\psi'(u_n) \to 0 \text{ in } (\mathcal{D}_0^{1,2}(\Omega))', \end{aligned}$$

then $\psi'(v) = 0$. *Moreover the sequence*

$$w_n(x) := u_n(x) - \lambda_n^{(2-N)/2} v\left(\frac{x - y_n}{\lambda_n}\right)$$

satisfies

$$\begin{aligned} &||w_n||^2 = ||u_n||^2 - ||v||^2 + o(1), \\ &\psi(w_n) \to c - \psi(v), \\ &\psi'(w_n) \to 0 \text{ in } (\mathcal{D}_0^{1,2}(\Omega))'. \end{aligned}$$

Proof. 1) Since $v_n \rightharpoonup v$ in $\mathcal{D}^{1,2}(\mathbb{R}^N)$, it is clear that

$$
\begin{aligned}
||w_n||^2 &= ||v_n - v||^2 \\
&= ||v_n||^2 - ||v||^2 + o(1) \\
&= ||u_n||^2 - ||v||^2 + o(1).
\end{aligned}
$$

2) According to the Brézis-Lieb lemma, we obtain

$$
\begin{aligned}
\psi(w_n) &= \psi(v_n - v) \\
&= \psi(v_n) - \psi(v) + o(1) \\
&= \psi(u_n) - \psi(v) + o(1) \\
&= c - \psi(v) + o(1).
\end{aligned}
$$

3) Since $v_n \rightharpoonup v$ in $\mathcal{D}^{1,2}(\mathbb{R}^N)$ and $\psi'(u_n) \to 0$ in $(\mathcal{D}_0^{1,2}(\Omega))'$, it is easy to verify, using (4), that $\psi'(v) = 0$. In particular $v \in \mathcal{C}^2(\mathbb{R}^N)$. Lemma 8.9 implies that, in $(\mathcal{D}_0^{1,2}(\Omega))'$,

$$
\begin{aligned}
\psi'(w_n) &= \psi'(u_n) - \psi'\left(\lambda_n^{(2-N)/2} v\left(\frac{\cdot - y_n}{\lambda_n}\right)\right) + o(1) \\
&= o(1). \qquad \square
\end{aligned}
$$

Lemma 8.12. *If $u \in \mathcal{D}^{1,2}(\mathbb{R}^N)$ and $v \in \mathcal{D}(\mathbb{R}^N)$ then*

$$
\int v^2 |u|^{2^*} \le S^{-1} \left(\int_{\text{supp}\, v} |u|^{2^*}\right)^{2/N} \int |\nabla(vu)|^2.
$$

Proof. The Hölder inequality implies that

$$
\int v^2 |u|^{2^*} \le \left(\int_{\text{supp}\, v} |u|^{2^*}\right)^{2/N} \left(\int |vu|^{2^*}\right)^{(N-2)/N}.
$$

It suffices then to use the Sobolev inequality. $\square$

The proof of the next result is inspired by the approach of [22].

Theorem 8.13. *(Struwe, 1984). Under assumption (B), let $(u_n) \subset \mathcal{D}_0^{1,2}(\Omega)$ be such that*

$$
\varphi(u_n) \to c, \varphi'(u_n) \to 0 \text{ in } (\mathcal{D}_0^{1,2}(\Omega))'.
$$

Then, replacing (u_n) if necessary by a subsequence, there exists a solution $v_0 \in \mathcal{D}_0^{1,2}(\Omega)$ of

$$
-\Delta u + a\, u = \lambda |u|^{2^*-2} u,
$$

a finite sequence $(v_1, ..., v_k) \subset \mathcal{D}^{1,2}(\mathbb{R}^N)$ of solutions of

$$
-\Delta u = \lambda |u|^{2^*-2} u,
$$

and k sequences (y_n^k), (λ_n^k) satisfying $\lambda_n^k > 0$, $y_n^k \in \Omega$ and

$$\frac{1}{\lambda_n^k} \, \mathrm{dist}(y_n^k, \partial\Omega) \to \infty, n \to \infty,$$

$$\left\| u_n - v_0 - \sum_{j=1}^k (\lambda_n^j)^{(2-N)/2} v_j \left(\frac{\cdot - y_n^j}{\lambda_n^j} \right) \right\| \to 0,$$

$$\| u_n \|^2 \to \sum_{j=0}^k \| v_j \|^2,$$

$$\varphi(v_0) + \sum_{j=1}^k \psi(v_j) = c.$$

Moreover, if $k \geq 1$ and $u_n \geq 0$ a.e. on Ω for all n, then $v_j > 0$ on $\mathbb{R}^N$ for $1 \leq j \leq k$.

Proof. 1) As in the proof of Lemma 1.20, (u_n) is bounded. We may assume that $u_n \rightharpoonup v_0$ in $\mathcal{D}_0^{1,2}(\Omega)$, $u_n \to v_0$ a.e. on Ω. By Lemma 8.10, $\varphi'(v_0) = 0$ and $u_n^1 := u_n - v_0$ is such that

(5)
$$\| u_n^1 \|^2 = \| u_n \|^2 - \| v_0 \|^2 + o(1),$$
$$\psi(u_n^1) \to c - \varphi(v_0),$$
$$\psi'(u_n^1) \to 0 \text{ in } (\mathcal{D}_0^{1,2}(\Omega))'.$$

2) If $u_n^1 \to 0$ in $L^{2^*}(\Omega)$, then, since $\psi'(u_n^1) \to 0$, it follows that $u_n^1 \to 0$ in $\mathcal{D}_0^{1,2}(\Omega)$ and the proof is complete. If $u_n^1 \not\to 0$ in $L^{2^*}(\Omega)$ we may assume that

$$\int |u_n^1|^{2^*} > \delta$$

for some $0 < \delta < (S/2\lambda)^{N/2}$. Let us define the Levy concentration function:

$$Q_n(r) := \sup_{y \in \mathbb{R}^N} \int_{B(y,r)} |u_n|^{2^*}.$$

Since $Q_n(0) = 0$ and $Q_n(\infty) > \delta$, there exists sequences (y_n^1), (λ_n^1) such that $y_n^1 \in \Omega$, $\lambda_n^1 > 0$ and

$$\delta = \sup_{y \in \mathbb{R}^N} \int_{B(y,\lambda_n^1)} |u_n^1|^{2^*} = \int_{B(y_n^1,\lambda_n^1)} |u_n^1|^{2^*}.$$

Let us define $v_n^1(x) := (\lambda_n^1)^{(N-2)/2} u_n^1(\lambda_n^1 x + y_n^1)$. We may assume that $v_n^1 \rightharpoonup v_1$ in $\mathcal{D}^{1,2}(\mathbb{R}^N)$, $v_n^1 \to v_1$ a.e. on $\mathbb{R}^N$. It is clear that

$$\delta = \sup_{y \in \mathbb{R}^N} \int_{B(y,1)} |v_n^1|^{2^*} = \int_{B(0,1)} |v_n^1|^{2^*}.$$

3) Let us define $\Omega_n := (1/\lambda_n^1)(\Omega - y_n^1)$ and let $f_n \in \mathcal{D}_0^{1,2}(\Omega)$ be such that

$$\forall h \in \mathcal{D}_0^{1,2}(\Omega), \langle \psi'(u_n^1), h \rangle = \int_\Omega \nabla f_n \nabla h.$$

Then $g_n(x) := (\lambda_n^1)^{(N-2)/2} f_n(\lambda_n^1 x + y_n^1)$ satisfies

$$\forall h \in \mathcal{D}_0^{1,2}(\Omega_n), \langle \psi'(v_n^1), h \rangle = \int_{\Omega_n} \nabla g_n \nabla h$$

and

$$\int_{\Omega_n} |\nabla g_n|^2 = \int_\Omega |\nabla f_n|^2 = o(1).$$

If $v_1 = 0$ then $v_n^1 \to 0$ in $L^2_{\text{loc}}(\mathbb{R}^N)$.

Let $h \in \mathcal{D}(\mathbb{R}^N)$ be such that supp $h \subset B(y,1)$ for some $y \in \mathbb{R}^N$. It follows then from the preceding lemma that

$$\begin{aligned}
\int_{\Omega_n} |\nabla(hv_n^1)|^2 &= \int_{\Omega_n} \nabla v_n^1 \nabla(h^2 v_n^1) + o(1) \\
&= \lambda \int h^2 |v_n^1|^{2^*} + \int \nabla g_n \nabla(h^2 v_n^1) + o(1) \\
&\leq \lambda S^{-1} \delta^{2/N} \int |\nabla(hv_n^1)|^2 + o(1) \\
&= \frac{1}{2} \int |\nabla(hv_n^1)|^2 + o(1).
\end{aligned}$$

We obtain $\nabla v_n^1 \to 0$ in $L^2_{\text{loc}}(\mathbb{R}^N)$ and $v_n^1 \to 0$ in $L^{2^*}_{\text{loc}}(\mathbb{R}^N)$. This is impossible since $\int_{B(0,1)} |v_n^1|^{2^*} = \delta > 0$. Thus we have proved that $v_1 \neq 0$.

4) Using the fact that Ω is bounded, we may assume that

$$\lambda_n^1 \to \lambda_0^1 \geq 0, \quad y_n^1 \to y_0^1 \in \bar{\Omega}.$$

If $\lambda_0^1 > 0$, the fact that $u_n^1 \to 0$ in $\mathcal{D}_0^{1,2}(\Omega)$ implies that $v_n^1 \to 0$ in $\mathcal{D}^{1,2}(\mathbb{R}^N)$. This is a contradiction.

If $\lambda_n^1 \to 0$ and

$$\lim_{n\to\infty} \frac{1}{\lambda_n^1} \text{dist}(y_n^1, \partial\Omega) < \infty,$$

we may assume that

$$\lim_{n\to\infty} \frac{1}{\lambda_n^1} \text{dist}(y_n^1, \partial\Omega) = d.$$

It is then easy to verify that v_1 in a solution of

$$-\Delta u = \lambda |u|^{2^*-2} u$$

in a half-space. By the Pohozaev identity, v_1 must vanish identically. This is also a contradiction. Hence

$$\lim_{n \to \infty} \frac{1}{\lambda_n^1} \operatorname{dist}(y_n^1, \partial\Omega) = \infty.$$

By (5) and Lemma 8.11, $\psi'(v_1) = 0$ and

$$u_n^2(x) := u_n^1(x) - (\lambda_n^1)^{(2-N)/2} v_1 \left(\frac{x - y_n^1}{\lambda_n^1} \right)$$

satisfies

$$\|u_n^2\| = \|u_n\|^2 - \|v_0\|^2 - \|v_1\|^2 + o(1),$$
$$\psi(u_n^2) \to c - \varphi(v_0) - \psi(v_1),$$
$$\psi'(u_n^2) \to 0 \text{ in } (\mathcal{D}_0^{1,2}(\Omega))'.$$

5) Any nontrivial critical point of ψ satisfies

$$S|u|_{2^*}^2 \le \|u\|^2 = \lambda|u|_{2^*}^{2^*}$$

so that

$$\psi(u) \ge c^* := \frac{\lambda}{N} \left(\frac{S}{\lambda} \right)^{N/2}.$$

Iterating the above procedure, we construct sequences (v_j), (λ_j^n) and (y_j^n). Since, for every j, $\psi(v_j) \ge c^*$, the iteration must terminate at some finite index.

6) If $u_n \ge 0$ a.e. on Ω for all n, it suffices to use the functionals

$$\varphi_+(u) := \int_\Omega \left[\frac{|\nabla u|^2}{2} + a\frac{u^2}{2} - \lambda\frac{u_+^{2^*}}{2^*} \right], u \in \mathcal{D}_0^{1,2}(\Omega),$$

$$\psi_+(u) := \int_{\mathbb{R}^N} \left[\frac{|\nabla u|^2}{2} - \lambda\frac{u_+^{2^*}}{2^*} \right], u \in \mathcal{D}^{1,2}(\mathbb{R}^N),$$

and the maximum principle. $\square$

We consider the following minimization problems

$$S_a := \inf_{\substack{u \in \mathcal{D}_0^{1,2}(\Omega) \\ |u|_{2^*}=1}} \int_\Omega [|\nabla u|^2 + au^2],$$

$$S := \inf_{\substack{u \in \mathcal{D}^{1,2}(\mathbb{R}^N) \\ |u|_{2^*}=1}} \int_{\mathbb{R}^N} |\nabla u|^2.$$

Theorem 8.14. (Brézis-Nirenberg, 1983). *Under assumption* (B), *if* $S_a < S$, *then every sequence* $(u_n) \subset \mathcal{D}_0^{1,2}(\Omega)$ *satisfying*

$$|u_n|_{2^*} = 1, \int_\Omega [|\nabla u_n|^2 + a u_n^2] \to S_a, n \to \infty,$$

contains a convergent subsequence. In particular, there exists a minimizer for S_a.

Proof. It suffices to modify the proof of Theorem 8.6. $\square$

8.4 Symmetric domains

When Ω and a are invariant be a group of orthogonal transformations, it is possible to improve the above theorem.

Let G be a subgroup of $\mathbf{O}(N)$. Let Ω be an invariant domain of $\mathbb{R}^N$ and let $a \in C^\infty(\Omega)$ be an invariant function satisfying (B).

As in Definition 1.23, the action of G on $\mathcal{D}_0^{1,2}(\Omega)$ is defined by

$$gu(x) := u(g^{-1}x).$$

The subspace of invariant functions is defined by

$$\mathcal{D}_{0,G}^{1,2}(\Omega) := \{u \in \mathcal{D}_0^{1,2}(\Omega) : gu = u, \forall g \in G\}.$$

We define also

$$m(G) \quad := \quad \inf_{|x|=1} \#\{gx : g \in G\},$$

$$S_G \quad := \quad \inf_{\substack{u \in \mathcal{D}_{0,G}^{1,2}(\Omega) \\ |u|_{2^*}=1}} \int_\Omega [|\nabla u|^2 + a\, u^2].$$

Theorem 8.15. *(P.L. Lions, 1985). Under assumption (B), if Ω and a are G-invariant, $0 \notin \Omega$ and $S_G < m(G)^{2/N} S$, then every sequence $(u_n) \subset \mathcal{D}_{0,G}^{1,2}(\Omega)$ satisfying*

$$|u_n|_{2^*} = 1, \int_\Omega [|\nabla u_n|^2 + a\, u_n^2] \to S_G, n \to \infty,$$

contains a convergent subsequence. In particular, there exists a minimizer for S_G.

Proof. It suffices to modify the proof of Theorem 8.8. $\square$

Appendix A :

Superposition operator

1. Domains with finite measure

In this appendix, we consider the continuity of the superposition operator

$$A : L^p(\Omega) \to L^q(\Omega) : u \mapsto f(x, u).$$

Lemma A.1. *Let Ω be an open subset of $\mathbb{R}^N$ and $1 \leq p < \infty$. If $v_n \to u$ in $L^p(\Omega)$, there exists a subsequence (w_n) of (v_n) and $g \in L^p(\Omega)$ such that, almost everywhere on Ω, $w_n(x) \to u(x)$ and*

$$|u(x)|, |w_n(x)| \leq g(x).$$

Proof. Going if necessary to a subsequence, we can assume that $v_n(x) \to u(x)$ a.e. on Ω. There exists a subsequence (w_n) of (v_n) such that

$$|w_{j+1} - w_j|_p \leq 2^{-j}, \quad \forall j \geq 1.$$

Let us define

$$g(x) := |w_1(x)| + \sum_{j=1}^{\infty} |w_{j+1}(x) - w_j(x)|.$$

It is clear that, a.e. on Ω, $|w_n(x)| \leq g(x)$ and so $|u(x)| \leq g(x)$. $\square$

Theorem A.2. *Assume that $|\Omega| < \infty$, $1 \leq p$, $r < \infty$, $f \in \mathcal{C}(\bar{\Omega} \times \mathbb{R})$ and*

$$|f(x, u)| \leq c(1 + |u|^{p/r}).$$

Then, for every $u \in L^p(\Omega)$, $f(., u) \in L^r(\Omega)$ and the operator

$$A : L^p(\Omega) \to L^r(\Omega) : u \mapsto f(x, u)$$

is continuous.

Proof. 1) Assume that $u \in L^p(\Omega)$. Since

$$|f(x, u)|^r \leq c^r(1 + |u|^{p/r})^r \in L^1(\Omega),$$

it follows that $f(., u) \in L^r(\Omega)$.

2) Assume that $u_n \to u$ in $L^p(\Omega)$. Consider a subsequence (v_n) of (u_n). Let (w_n) and g be given by the preceding lemma. Since

$$|f(x, w_n) - f(x, u)|^r \leq 2^r c^r(1 + |g|^{p/r})^r \in L^1(\Omega),$$

it follows from Lebesgue dominated convergence theorem that $Aw_n \to Au$ in $L^r(\Omega)$. But then $Au_n \to Au$ in $L^2(\Omega)$. $\square$

2. Domains with infinite measure

Definition A.3. *On the space $L^p(\Omega) \cap L^q(\Omega)$, we define the norm*

$$|u|_{p \wedge q} := |u|_p + |u|_q.$$

On the space $L^p(\Omega) + L^q(\Omega)$, we define the norm

$$|u|_{p \vee q} = \inf\{|v|_p + |w|_q : v \in L^p(\Omega), w \in L^q(\Omega), u = v + w\}.$$

Theorem A.4. *Assume that $1 \leq p, q, r, s < \infty$, $f \in \mathcal{C}(\bar{\Omega} \times \mathbb{R})$ and*

$$|f(x, u)| \leq c(|u|^{p/r} + |u|^{q/s}).$$

Then, for every $u \in L^p(\Omega) \cap L^q(\Omega)$, $f(., u) \in L^r(\Omega) + L^s(\Omega)$ and the operator

$$A : L^p(\Omega) \cap L^q(\Omega) \to L^r(\Omega) + L^s(\Omega) : u \mapsto f(x, u)$$

is continuous.

Proof. Let $\psi \in \mathcal{D}(]-2, 2[)$ be such that $\psi = 1$ on $]-1, 1[$ and define

$$g(x, u) := \psi(u)f(x, u), h(x, u) := (1 - \psi(u))f(x, u).$$

We can assume that $p/r \leq q/s$. Hence we obtain

$$|g(x, u)| \leq a|u|^{p/r}, \quad |h(x, u)| \leq b|u|^{q/s}.$$

Assume that $u_n \to u$ in $L^p \cap L^q$. As in the proof of Theorem A.2, we have that $g(x, u_n) \to g(x, u)$ in L^r and $h(x, u_n) \to h(x, u)$ in L^s. Since

$$|f(x, u_n) - f(x, u)|_{r \vee s} \leq |g(x, u_n) - g(x, u)|_r + |h(x, u_n) - h(x, u)|_s$$

it follows that $f(x, u_n) \to f(x, u)$ in $L^r + L^s$. $\square$

Appendix B :

Variational identities

1. Virial theorem

We give a formal argument explaining some variational identities. Let X be a Banach space and let $T(t) : X \to X$ be a family of transformations such that

$$T(1) = \mathrm{id} \, .$$

If $u \in X$ is a critical point of $\varphi \in C^1(X, \mathbb{R})$, we conjecture that

$$\frac{\partial}{\partial t}\Big|_{t=1} \varphi(T(t)u) = 0.$$

Consider, for example, the following situation

$$X := \mathcal{D}^{1,2}(\mathbb{R}^N),$$

$$T(t)u(x) := u(x/t), \quad t > 0,$$

$$\varphi(u) := \int_{\mathbb{R}^N} \left[\frac{|\nabla u|^2}{2} - F(u) \right].$$

We have that

$$\varphi(T(t)u) = \frac{t^{N-2}}{2} \int_{\mathbb{R}^N} |\nabla u|^2 - t^N \int_{\mathbb{R}^N} F(u)$$

and

$$\frac{\partial}{\partial t}\Big|_{t=1} \varphi(T(t)u) = \frac{N-2}{2} \int_{\mathbb{R}^N} |\nabla u|^2 - N \int_{\mathbb{R}^N} F(u).$$

If u is a critical point of φ, we conjecture that

$$\frac{N-2}{2} \int_{\mathbb{R}^N} |\nabla u|^2 = N \int_{\mathbb{R}^N} F(u).$$

2. Bounded domains

In this section, we consider the problem

$(\mathcal{P}_1)$
$$\begin{cases} -\Delta u = f(u), \\ u \in H_0^1(\Omega), \end{cases}$$

where $f \in \mathcal{C}^1(\mathbb{R}, \mathbb{R})$ and Ω is a smooth bounded domlain of $\mathbb{R}^N$, $N \geq 3$.
The action of $t > 0$ is defined by

$$T(t)u(x) := u(x/t)$$

and the corresponding generator is

$$\frac{\partial}{\partial t}\Big|_{t=1} T(t) = -x \cdot \nabla.$$

In order to prove the Pohozaev identity, we multiply $-\Delta u = f(u)$ by $x \cdot \nabla u$ and we integrate by parts. We define

$$F(u) := \int_0^u f(s)ds.$$

Theorem B.1. (Pohozaev identity, 1965). *Let $u \in H_{\text{loc}}^2(\bar{\Omega})$ be a solution of $(\mathcal{P}_1)$ such that $F(u) \in L^1(\Omega)$. Then u satisfies*

$$\frac{1}{2}\int_{\partial\Omega} |\nabla u|^2 \sigma \cdot \nu d\sigma = N \int_\Omega F(u)dx - \frac{N-2}{2}\int_\Omega |\nabla u|^2 dx,$$

where ν denotes the unit outward normal to $\partial\Omega$.

Proof. It follows from $(\mathcal{P}_1)$ that

$$0 = (\Delta u + f(u))x \cdot \nabla u.$$

It is clear that

$$f(u)\, x \cdot \nabla u \;=\; \operatorname{div}(xF(u)) - NF(u),$$
$$\Delta u\, x \cdot \nabla u \;=\; \operatorname{div}(\nabla u\, x \cdot \nabla u) - |\nabla u|^2 - x \cdot \nabla\left(\frac{|\nabla u|^2}{2}\right)$$
$$=\; \operatorname{div}\left(\nabla u\, x \cdot \nabla u - x\frac{|\nabla u|^2}{2}\right) + \frac{N-2}{2}|\nabla u|^2.$$

Integrating by parts, we obtain

$$\int_{\partial\Omega}\left[\sigma F(u) + \nabla u\, \sigma \cdot \nabla u - \sigma\frac{|\nabla u|^2}{2}\right] \cdot \nu d\sigma = \int_\Omega \left[NF(u) - \frac{N-2}{2}|\nabla u|^2\right] dx.$$

But, on $\partial\Omega$, $u = 0$ so that

$$F(u) = 0, \quad \nabla u = \nabla u \cdot \nu\nu$$

and this implies

$$\frac{1}{2}\int_{\partial\Omega} |\nabla u|^2 \sigma \cdot \nu d\sigma = \int_{\Omega} \left[NF(u) - \frac{N-2}{2}|\nabla u|^2 \right] dx. \qquad \square$$

Corollary B.2. (Rellich identity, 1940). *Let* $u \in H^2_{loc}(\bar{\Omega})$ *be a solution of*

$$\begin{cases} -\Delta u = \lambda u, \\ u \in H^1_0(\Omega). \end{cases}$$

Then u *satisfies*

$$\frac{1}{2}\int_{\partial\Omega} |\nabla u|^2 \sigma \cdot \nu d\sigma = \int_{\Omega} \lambda u^2 dx.$$

Proof. It suffices to use the Pohozaev identity since

$$\int_{\Omega} F(u)dx = \frac{\lambda}{2}\int_{\Omega} u^2 dx, \int_{\Omega} |\nabla u|^2 dx = \lambda \int_{\Omega} u^2 dx. \qquad \square$$

Generalizations of the Pohozaev identity are proved in [65] and [87].

3. Unbounded domains

We now consider the problem

$$(\mathcal{P}_2) \qquad \begin{cases} -\Delta u = f(u), \\ u \in \mathcal{D}^{1,2}_0(\Omega), \end{cases}$$

where $f \in C^1(\mathbb{R}, \mathbb{R})$, $f(0) = 0$, and Ω is a smooth unbounded domain of $\mathbb{R}^N$, $N \geq 3$.

In order to prove the Pohozaev identity, we use a truncation argument due to Kavian.

Theorem B.3. *Let* $u \in H^q_{loc}(\bar{\Omega})$ *be a solution of* $(\mathcal{P}_2)$ *such that* $F(u) \in L^1(\Omega)$. *Then the Pohozaev identity is valid.*

Proof. Let $\psi \in \mathcal{D}(\mathbb{R})$ be such that $0 \leq \psi \leq 1$, $\psi(r) = 1$ for $r \leq 1$ and $\psi(r) = 0$ for $r \geq 2$. Define on $\mathbb{R}^N$

$$\psi_n(x) := \psi(|x|^2/n^2).$$

There exists $c \geq 0$ such that, for every n,

$$\psi_n \leq c, \quad |x| |\nabla \psi_n(x)| \leq c.$$

It follows from $(\mathcal{P}_2)$ that, for every n,

$$0 = (\Delta u + f(u))\psi_n \, x \cdot \nabla u.$$

It is clear that, for every n,

$$\psi_n f(u) \, x \cdot \nabla u = \operatorname{div}(x \psi_n F(u)) - N \psi_n F(u) - F(u) \, x \cdot \nabla \psi_n, \ \psi_n \Delta u \, x \cdot \nabla u$$

$$= \operatorname{div}(\nabla u \, \psi_n \, x \cdot \nabla u) - \psi_n |\nabla u|^2 - \psi_n \, x \cdot \nabla \left(\frac{|\nabla u|^2}{2} \right) - x \cdot \nabla u \, \nabla \psi_n \cdot \nabla u$$

$$= \operatorname{div}([\nabla u \, x \cdot \nabla u - x \tfrac{|\nabla u|^2}{2}]\psi_n) + \tfrac{N-2}{2} \psi_n |\nabla u|^2 + \tfrac{|\nabla u|^2}{2} x \cdot \nabla \psi_n - x \cdot \nabla u \, \nabla \psi_n \cdot \nabla u.$$

Integrating by parts, we obtain, for every n,

$$\int_{\partial \Omega} \left[\sigma F(u) + \nabla u \, \sigma \cdot \nabla u - \sigma \frac{|\nabla u|^2}{2} \right] \psi_n \cdot \nu d\sigma$$

$$= \int_\Omega \left(\left[N F(u) - \frac{N-2}{2} |\nabla u|^2 \right] \psi_n + F(u) \, x \cdot \nabla \psi_n \right.$$

$$\left. + \frac{|\nabla u|^2}{2} x \cdot \nabla \psi_n - x \cdot \nabla u \, \nabla \psi_n \cdot \nabla u \right) dx.$$

The Lebesgue dominated convergence theorem implies that

$$\int_{\partial \Omega} \left[\sigma F(u) + \nabla u \, \sigma \cdot \nabla u - \sigma \frac{|\nabla u|^2}{2} \right] \cdot \nu d\sigma = \int_\Omega \left[N F(u) - \frac{N-2}{2} \frac{|\nabla u|^2}{2} \right] dx.$$

It is then easy to conclude as in the proof of Theorem B.1. $\square$

Corollary B.4. Let $f \in \mathcal{C}^1(\mathbb{R}, \mathbb{R})$ be such that $f(0) = 0$ and let $u \in H^2_{\text{loc}}(\mathbb{R}^N)$ be a solution of

$$\begin{cases} -\Delta u = f(u), \\ u \in \mathcal{D}^{1,2}(\mathbb{R}^N), \end{cases}$$

such that $F(u) \in L^1(\mathbb{R}^N)$. Then u satisfies

$$\frac{N-2}{2} \int_{\mathbb{R}^N} |\nabla u|^2 = N \int_{\mathbb{R}^N} F(u).$$

Appendix C :

Symmetry of minimizers

1. Unconstrained problems

In this appendix, we study the symmetry of minimizers by the method of Orlando Lopes. This elementary but powerful method is applicable to systems. The results of this appendix are due to Lopes.

Let Ω be an open subset of $\mathbb{R}^N$, $2 \leq p \leq 2^*$, define

$$X_p := \left(\mathcal{D}_0^{1,2}(\Omega) \cap L^p(\Omega) \right)^M,$$

and consider the problem

$(\mathcal{P}_1)$
$$\begin{cases} \text{minimize } \int_\Omega \left[\frac{1}{2}|\nabla u|^2 - F(x,u) \right] dx, \\ u \in X_p. \end{cases}$$

If $|\Omega| < \infty$, it is clear that $X_p = \left(H_0^1(\Omega) \right)^M$. Our assumptions are the following:

(A1) Ω is symmetric with respect to the hyperplane $x_1 = 0$.

(A2) $F \in C^2(\Omega \times \mathbb{R}^M)$ and

$$F(-x_1, x_2, ..., x_N, u) = F(x_1, x_2, ..., x_N, u).$$

(A3) If $|\Omega| < \infty$ then, for some $c > 0$,

$$|\partial_u F(x,u)| \leq c(1 + |u|^{2^*-1}).$$

If $|\Omega| = \infty$ then $F(x,0) \equiv 0$ and, for some $c > 0$,

$$|\partial_u F(x,u)| \leq c(|u|^{p-1} + |u|^{2^*-1}).$$

Theorem C.1. *Under assumptions (A1-2-3), any solution of $(\mathcal{P}_1)$ is symmetric with respect to the hyperplane $x_1 = 0$.*

Proof. Define

$$\Omega_- := \{x \in \Omega : x_1 \leq 0\}, \qquad \Omega_+ := \{x \in \Omega : x_1 \geq 0\},$$
$$C_\pm := \int_{\Omega_\pm} \left[\frac{1}{2}|\nabla u|^2 - F(x, u)\right] dx.$$

Let v by the reflection with respect to the hyperplane $x_1 = 0$ of u restricted to Ω_+. Since $v \in X_p$, we have

$$C_+ + C_- \leq 2C_+.$$

Similarly, we obtain

$$C_+ + C_- \leq 2C_-.$$

Hence $C_+ = C_-$ and v is also a solution of $(\mathcal{P}_1)$. In particular, u and v satisfy the Euler systems

$$\begin{aligned} -\Delta u &= \partial_u F(x, u), \\ -\Delta v &= \partial_u F(x, v) \end{aligned}$$

and $u, v \in L^\infty(\Omega)$. Thus the vector field $w := u - v$ satisfies the linear system

$$-\Delta w = L(x)w,$$

where

$$L(x) := \int_0^1 \partial_u^2 F(x, u + t(v - u)) dt.$$

Since $w = 0$ on Ω_+, we conclude from the unique continuation principle that $w = 0$ on Ω. $\square$

The following result is applicable when Ω is $\mathbb{R}^N$ or a ball or an annulus or the exterior of a ball.

Theorem C.2. *If $F \in C^2(\Omega \times \mathbb{R}^M)$ satisfies (A3) and if, for every $g \in \mathbf{O}(N)$,*
a) $g\Omega = \Omega$,
b) $F(gx, u) \equiv F(x, u)$,
then any solution of $(\mathcal{P}_1)$ is radially symmetric.

Proof. By the preceding theorem, a solution of $(\mathcal{P}_1)$ is symmetric with respect to any hyperplane through the origin. $\square$

2. Constrained problems

Let $2 \leq p \leq 2^*$, define as before

$$X_p := \left(\mathcal{D}_0^{1,2}(\Omega) \cap L^p(\Omega) \right)^M$$

and consider the problem

$(\mathcal{P}_2)$ $\quad \begin{cases} \text{minimize } \int_\Omega \left[\frac{1}{2}|\nabla u|^2 - F(x,u) \right] dx, \\ \int_\Omega G(x,u) dx = 1, u \in X_p. \end{cases}$

Our assumptions are now the following:

(B1) Ω is invariant with respect to any translation in the x_1 direction.

(B2) $F, G \in C^2(\Omega \times \mathbb{R}^M)$ and F and G are independent of x_1.

(B3) $F(x,0) = G(x,0) \equiv 0$ and, for some $c > 0$,

$$|\partial_u F(x,u)| + |\partial_u G(x,u)| \leq c(|u|^{p-1} + |u|^{2^*-1}).$$

(B4) If $\int_\Omega G(x,u) dx = 1$ then $\partial_u G(x,u) \not\equiv 0$.

Theorem C.3. *Under assumptions (B1-2-3-4), any solution of $(\mathcal{P}_2)$ is symmetric with respect to the hyperplane $x_1 = 0$ after a translation in the x_1 direction.*

Proof. Define

$$\Omega_- := \{x \in \Omega : x_1 \leq 0\}, \qquad \Omega_+ := \{x \in \Omega : x_1 \geq 0\},$$
$$D_\pm := \int_{\Omega_\pm} G(x,u) dx$$

and

$$C_\pm := \int_{\Omega_\pm} \left[\frac{1}{2}|\nabla u|^2 - F(x,u) \right] dx.$$

After a translation in the x_1 direction, we can assume that $D_+ = D_- = 1/2$. Let v by the reflection with respect to the hyperplane $x_1 = 0$ of u restricted to Ω_+. Since v is admissible, we have

$$C_+ + C_- \leq 2C_+.$$

Similarly we obtain

$$C_+ + C_- \leq 2C_-.$$

Hence $C_+ = C_-$ and v is also of $(\mathcal{P}_2)$. By the Lagrange multiplier rule, there exists $\lambda, \tilde{\lambda} \in \mathbb{R}$ such that

$$- \Delta u - \partial_u F(x,u) = \lambda \partial_u G(x,u),$$
$$- \Delta v - \partial_u F(x,v) = \tilde{\lambda} \partial_u G(x,v)$$

and $u, v \in L^\infty(\Omega)$. Since $u = v$ on Ω_+, it follows from (B4) that $\lambda = \tilde{\lambda}$. Thus the vector field $w := u - v$ satisfies the linear system

$$-\Delta w = L(x)w$$

where

$$L(x) := \int_0^1 [\partial_u^2 F(x, u + t(v - u)) + \lambda \partial_u^2 G(x, u + t(v - u))]dt.$$

Since $w = 0$ on Ω_+, we conclude from the unique continuation principle that $w = 0$ on Ω. $\square$

Finally we consider the problem

$(\mathcal{P}_3)$ $\qquad \begin{cases} \text{minimize } \int_{\mathbb{R}^N} \left[\frac{1}{2}|\nabla u|^2 - F(u)\right]dx, \\ \int_{\mathbb{R}^N} G(u)dx = 1, u \in X_p. \end{cases}$

Theorem C.4. If $F, G \in C^2(\mathbb{R}^M)$ satisfy (B3) and (B4), then any solution of $(\mathcal{P}_3)$ is radially symmetric after a translation.

Proof. By the preceding theorem, a solution u of $(\mathcal{P}_3)$ is symmetric with respect to the hyperplanes $x_k = 0$, $k = 1, ..., N$, after a translation. In particular u is symmetric with respect to 0. Let π be a hyperplane through the origin. The preceding theorem implies the existence of an hyperplane $\tilde{\pi}$ parallel to π such that u is symmetric with respect to $\tilde{\pi}$. If $\tilde{\pi} \neq \pi$, it is easy to verify that u is periodic in the direction perpendicular to π using the fact that u is symmetric with respect to 0. Since $u \in L^p(\mathbb{R}^N)$, it follows that $u = 0$. This is a contradiction since $\int_{\mathbb{R}^N} G(u)dx = 1$. Thus u is symmetric with respect to π. Since u is symmetric with respect to any hyperplane through the origin, u is radially symmetric. $\square$

Example C.5. (Brézis-Nirenberg, 1983). Let Ω be an annulus of $\mathbb{R}^N$, $N \geq 3$, and define, for $2 \leq p \leq 2^*$,

$$S_p := \inf_{\substack{u \in H_0^1(\Omega) \\ |u|_p = 1}} |\nabla u|_2^2, \qquad \Sigma_p := \inf_{\substack{u \in H_{0,O(N)}^1(\Omega) \\ |u|_p = 1}} |\nabla u|_2^2$$

where $H_{0,O(N)}^1(\Omega) := \{u \in H_0^1(\Omega) : u \text{ is radially symmetric}\}$. By Proposition 1.43, S_{2^*} is never achieved. Lemma 4.5 implies that, for $1 \leq p \leq \infty$, the imbeddings

$$H_{0,O(N)}^1(\Omega) \subset L^p(\Omega)$$

are compact. In particular, Σ_{2^*} is always achieved. It follows that $S_{2^*} < \Sigma_{2^*}$ and, by continuity, for $2 < p < 2^*$ sufficiently close to 2^*,

$$S_p < \Sigma_p.$$

By Rellich theorem, S_p is achieved. Hence the solutions of the problem

$$\begin{cases} \text{minimize } \int_\Omega |\nabla u|^2 dx, \\ \int_\Omega |u|^p dx = 1, u \in H_0^1(\Omega) \end{cases}$$

are nonradial for $2 < p < 2^*$ sufficiently close to 2^*.

Appendix D :

Topological degree

1. Sard theorem

We will use Sard's theorem in the construction of degree theory.

Definition D.1. *Let X, Y be Banach spaces and let U be an open subset of X. A point $y \in Y$ is a singular value of $f \in \mathcal{C}^1(U, Y)$ if there exists $u \in U$ such that $y = f(u)$ and $f'(u)$ is not surjective. A point $y \in Y$ is a regular value of f if it is not a singular value.*

Theorem D.2. (Sard, 1942). *Let $U \subset \mathbb{R}^M$ and let $f \in \mathcal{C}^k(U, \mathbb{R}^N)$. If*

$$k > \max(0, M - N)$$

then the set of singular values of f has 0 measure in $\mathbb{R}^N$.

2. Topological degree

Let $U \subset \mathbb{R}^N$ be open and bounded. The topological degree of $f \in \mathcal{C}(\bar{U}, \mathbb{R}^N)$ represents in the "generic" case the "algebraic" number of zeros of f in U.

Definition D.3. *Let $f \in \mathcal{C}(\bar{U}, \mathbb{R}^N) \cap \mathcal{C}^2(U, \mathbb{R}^N)$ be such that $0 \notin f(\partial U)$ and 0 is a regular value of $f\big|_U$. If $u \in f^{-1}(0)$, then $f'(u)$ is invertible. By the inverse function theorem, $f^{-1}(0)$ is finite. The degree is defined by*

$$\deg(f, U) := \sum_{u \in f^{-1}(0)} \operatorname{sign} \det f'(u).$$

Lemma D.4. *Let $f \in \mathcal{C}(\bar{U}, \mathbb{R}^N) \cap \mathcal{C}^2(U, \mathbb{R}^N)$ be such that $0 \notin f(\partial U)$ and 0 is a regular value of $f\big|_U$. There exists an open neighborhood V of 0 in $\mathbb{R}^N$ such that every $y \in V$ is a regular value of $f\big|_U$ and*

$$\deg(f - y, U) = \deg(f, U).$$

Proof. As in the preceding definition, $f^{-1}(0) = \{u_1, \ldots, u_k\}$ and $f'(u_j)$ is invertible, $j = 1, \ldots, k$. By the inverse function theorem, there exists pairwise disjoint open neighborhoods $U_1, \ldots, U_k$ of $u_1, \ldots, u_k$ which are diffeomorphic to an open neighborhood W of 0 in $\mathbb{R}^N$. We choose

$$V := W \backslash f\left(\bar{U} \backslash \bigcup_{j=1}^{k} U_j\right). \qquad \square$$

The following lemma, proved by Nagumo in 1952, is our basic tool to extend topological degree.

Lemma D.5. *Let $h \in C([0,1] \times \bar{U}, \mathbb{R}^N) \cap C^2([0,1] \times U, \mathbb{R}^N)$ be such that*
a) $0 \notin h([0,1] \times \partial U)$,
b) 0 is a regular value of $h(0, .)\big|_U$ and $h(1, .)\big|_U$.
Then $\deg(h(0, .), U) = \deg(h(1, .)U)$.

Proof. We will use a transversality argument. Let

$$r := \min_{\substack{t \in [0,1] \\ u \in \partial U}} |h(t, u)| > 0.$$

Sard's theorem and the preceding lemma imply the existence of a regular value $y \in B(0, r)$ of $h\big|_{[0,1] \times U}$ such that

$$\deg(h(j, .), U) = \deg(h(j, .) - y, U), \quad j = 1, 2.$$

By the implicit function theorem, $h^{-1}(y)$ is a compact one-dimensional manifold with boundary contained in $[0,1] \times U$. In particular, $h^{-1}(y)$ consists of at most a finite number of curves $C_1, \ldots, C_k$.

For each $C := C_j$, we define a parametrization by arclength $v(s) := (u(s), t(s))$ and an orientation so that

$$h'(v)\dot{v} = 0, \quad |\dot{v}|^2 = 1, \quad \det\begin{pmatrix} h'(v) \\ \dot{v} \end{pmatrix} > 0.$$

Since

$$\dot{t} \det\begin{pmatrix} h'(v) \\ \dot{v} \end{pmatrix} = \det\begin{pmatrix} \partial_u h(t, u) & \partial_t h(t, u)\dot{t} \\ \dot{u} & \dot{t}^2 \end{pmatrix}$$

$$= \det\begin{pmatrix} \partial_u h(t, u) & h'(v)\dot{v} \\ \dot{u} & |\dot{v}|^2 \end{pmatrix} = \det \partial_u h(t, u)$$

we obtain

$$\operatorname{sign} \dot{t} = \operatorname{sign} \det \partial_u h(t, u).$$

We have four possible cases for each C_j:

a) C_j is a closed curve contained in $]0,1[\times U$,

b) C_j runs from one of the two hyperplanes $t = 0$ and $t = 1$ to the same one.

c) C_j runs from the hyperplane $t = 0$ to the hyperplane $t = 1$.

d) C_j runs from the hyperplane $t = 1$ to the hyperplane $t = 0$.

Let p be the number of curves of the case c) and let q be the number of curves of the case d). Then we have

$$d(h(0,.),U) = p - q = d(h(1,.),U). \qquad \square$$

Lemma D.6. Let $f \in \mathcal{C}(\bar{U},\mathbb{R}^N)$ be such that $0 \notin f(\partial U)$ and let $g_0, g_1 \in \mathcal{C}(\bar{U},\mathbb{R}^N) \cap \mathcal{C}^2(U,\mathbb{R}^N)$ be such that

$$\max_{u \in \partial U} |f(u) - g_j(u)| < \min_{u \in \partial U} |f(u)|, \quad j = 0,1$$

and 0 is a regular value of $g_0 \big|_U$ and $g_1 \big|_U$. Then $\deg(g_1,U) = \deg(g_2,U)$.

Proof. It suffices to define on $[0,1] \times \bar{U}$ the homotopy

$$h(t,u) := (1-t)g_0(u) + t\,g_1(u)$$

and to use the preceding lemma. $\square$

Definition D.7. Let $f \in \mathcal{C}(\bar{U},\mathbb{R}^N)$ be such that $0 \notin f(\partial U)$. The Weierstrass approximation theorem and Sard's theorem imply the existence of $g \in \mathcal{C}(\bar{U},\mathbb{R}^N) \cap \mathcal{C}^2(U,\mathbb{R}^N)$ such that 0 is a regular value of $g\big|_U$ and

$$\max_{u \in \partial U} |f(u) - g(u)| < \min_{u \in \partial U} |f(u)|.$$

The topological degree

$$\deg(f,U) := \deg(g,U)$$

is well defined by the preceding lemma.

Let us prove the basic properties of degree.

Theorem D.8. Let $f \in \mathcal{C}(\bar{U},\mathbb{R}^N)$ be such that $0 \notin f(\partial U)$.

a) *(Existence property).* If $\deg(f,U) \neq 0$ then $0 \in f(U)$.

b) *(Excision).* If V is an open subset of U such that $0 \notin f(\bar{V})$ then

$$\deg(f,U\backslash\bar{V}) = \deg(f,U).$$

Proof. Let (g_n) be a sequence of maps of $\mathcal{C}(\bar{U},\mathbb{R}^N) \cap \mathcal{C}^2(U,\mathbb{R}^N)$ converging uniformly to f on $\bar{U}$ and such that 0 is always a regular value of $g_n\big|_U$. For n large enough, we have $\deg(f,U) = \deg(g_n,U)$.

If $\deg(f, U) = 1$, there exists $u_n \in U$ such that $g_n(u_n) = 0$. We can assume that $u_n \to u \in \bar{U}$ and so $f(u) = 0$. But $0 \notin f(\partial U)$, so that $0 \in f(U)$.

If $0 \notin f(\bar{V})$, then for n large enough

$$\deg(f, U) = \deg(g_n, U) = \deg(g_n, U \backslash \bar{V}) = \deg(f, U \backslash \bar{V}). \qquad \square$$

Theorem D.9. (Homotopy invariance). *Let* $h \in \mathcal{C}([0,1] \times \bar{U}, \mathbb{R}^N)$ *be such that* $0 \notin h([0,1] \times \partial U)$. *Then*

$$\deg(h(0, .), U) = \deg(h(1, .), U).$$

Proof. The Weierstrass approximation theorem and Sard's theorem imply the existence of $g \in \mathcal{C}([0,1] \times \bar{U}, \mathbb{R}^N) \cap \mathcal{C}^2([0,1] \times U, \mathbb{R}^N)$ such that

$$\max_{\substack{t \in [0,1] \\ u \in \partial U}} |h(t, u) - g(t, u)| < \min_{\substack{t \in [0,1] \\ u \in \partial U}} |h(t, u)|$$

and 0 is a regular value of $g(0, .)\big|_U$ and $g(1, .)\big|_U$. One concludes by using Lemma D.5 and the definition of degree. $\square$

3. Non retractability theorem

We define

$$B^N := \{x \in \mathbb{R}^N : |x| \leq 1\},$$
$$S^{N-1} := \{x \in \mathbb{R}^N : |x| = 1\}.$$

Definition D.10. *A retraction from a topological space* X *to a subspace* Y *is a continuous map* $r : X \to Y$ *such that* $r(y) = y$ *for every* $y \in Y$.

Theorem D.11. *There is no retraction from* B^N *to* S^{N-1}.

Proof. Assume, by contradiction, that $r : B^N \to S^{N-1}$ is a retraction and let U be the interior of B^N. Using the homotopy

$$h(t, u) := (1 - t)u + t\, r(u),$$

we deduce

$$\deg(r, U) = \deg(\mathrm{id}, U) = 1.$$

We obtain, by existence property, the contradiction

$$0 \in r(U) \subset S^{N-1}. \qquad \square$$

Corollary D.12. (Brouwer fixed point theorem). *Any continuous map from* B^N *to* B^N *has a fixed point.*

Proof. Assume, by contradiction, that $f \in \mathcal{C}(B^N, B^N)$ has no fixed point. Denote by $r(u)$ the intersection of S^{N-1} and the half line from $f(u)$ to u. Then r is a retraction from B^N to S^{N-1}. $\square$

Remarks D.13. a) The non retractability theorem follows also from Brouwer fixed point theorem. Assume that r is a retraction from B^N to S^{N-1}. Then $f := -r$ is a continuous map from B^N to B^N without fixed point.

b) It is easy to prove directly Brouwer fixed point theorem by using degree theory.

Dugundji proved in 1951 that there is no generalization of Brouwer fixed point theorem to infinite dimensional normed spaces.

Example D.14. (Kakutani, 1943). Let B be the unit ball of

$$\ell^2(\mathbb{N}) := \{(x_k) \in \mathbb{R}^N : \sum_{k \in \mathbb{N}} x_k^2 < \infty\}$$

with the norm

$$|(x_k)| := \left(\sum_{k \in \mathbb{N}} x_k^2\right)^{1/2}.$$

The map $f : B \to B$ defined by

$$f(x_0, x_1, \ldots) := \left((1 - \sum_{k \in \mathbb{N}} x_k^2)^{1/2}, x_0, x_1, \ldots\right)$$

has no fixed point.

4. Borsuk-Ulam theorem

In order to compute the degree of odd maps, we need the following lemma.

We give the proof of Gromes.

Lemma D.15. *Let U be an open bounded symmetric neighborhood of 0 in $\mathbb{R}^N$ and let $f : \bar{U} \to \mathbb{R}^N$ be continuous and odd. For every $\varepsilon > 0$, there exists $g \in \mathcal{C}(\bar{U}, \mathbb{R}^N) \cap \mathcal{C}^1(U, \mathbb{R}^N)$ such that*
a) g is odd,
b) 0 is a regular value of $g|_U$,
c) $\max_{u \in \bar{U}} |f(u) - g(u)| < \varepsilon$.

Proof. 1) Assume that V is an open subset of $\mathbb{R}^N$ and that $F : V \to \mathbb{R}^N$, $G : V \to \mathbb{R} \setminus \{0\}$ are differentiable. It is easy to verify that $y \in \mathbb{R}^N$ is a regular value of F/G if and only if 0 is a regular value of $H_y := F - Gy$. By Sard theorem, 0 is a regular value of H_y for almost every $y \in \mathbb{R}^N$.

2) By the Weierstrass approximation theorem, there exists $F \in \mathcal{C}(\bar{U}, \mathbb{R}^N) \cap \mathcal{C}^1(U, \mathbb{R}^N)$ such that

$$\max_{u \in \bar{U}} |f(u) - F(u)| < \eta := \varepsilon/(N+1).$$

By replacing $F(u)$ by $\frac{1}{2}(F(u) - F(-u))$, we can assume that F is odd. Moreover by replacing $F(u)$ by $F(u) - \lambda u$ where λ is small, we can assume that $F'(0)$ is invertible.

3) We define

$$E_k := \{x \in \mathbb{R}^N : x_k = 0\}, U_k := U \backslash (E_1 \cap \ldots \cap E_k).$$

We choose $y^1 \in \mathbb{R}^N$ so that 0 is regular value of $h_1(u) := F(u) - u_1^3 y^1$ restricted to $U \backslash E_1$ and

$$\max_{u \in \bar{U}} |F(u) - h_1(u)| < \eta.$$

If h_k is defined, we choose $y^{k+1} \in \mathbb{R}^N$ so that 0 is a regular value of $h_{k+1}(u) := h_k(u) - u_{k+1}^3 y^{k+1}$ restricted to $U \backslash E_{k+1}$ and

$$\max_{u \in \bar{U}} |h_k(u) - h_{k+1}(u)| < \eta.$$

Since $U_{k+1} = (U \backslash E_{k+1}) \cup (U_k \cap E_{k+1})$, 0 is also a regular value of h_{k+1} restricted to U_{k+1}. Finally the map g is given by h_N. $\square$

Theorem D.16. (Borsuk theorem). *Under the assumptions of the preceding lemma, if $0 \notin f(\partial U)$ then $\deg(f, U)$ is odd.*

Proof. Let g be given by the preceding lemma where $\varepsilon := \min_{u \in \partial U} |f(u)|$. We obtain by definition

$$\deg(f, U) = \deg(g, U) = 1, \bmod 2. \qquad\qquad \square$$

Theorem D.17. (Borsuk-Ulam theorem). *Let U be an open bounded symmetric neighborhood of 0 in $\mathbb{R}^N$. Every continuous odd map $f : \partial U \to \mathbb{R}^{N-1}$ has a zero.*

Proof. Assume, by contradiction, that $f : \partial U \to \mathbb{R}^{N-1} \backslash \{0\}$ is a continuous odd map. By the Tietze theorem, f has a continuous extension $g : \bar{U} \to \mathbb{R}^{N-1}$. Moreover we can assume that g is odd. By the preceding theorem, $\deg(g, U) \neq 0$. Define $r := \min_{u \in \partial U} |g(u)| > 0$. We obtain, by homotopy invariance and existence property, the contradiction

$$B(0, r) \subset g(\bar{U}) \subset \mathbb{R}^{N-1}. \qquad\qquad \square$$

Remark D.18. The non retractability theorem follows from Borsuk-Ulam theorem. Suppose that r is a retraction from B^{N-1} to S^{N-2}. We can assume that r is odd. Then the map $f : S^{N-1} \to S^{N-2}$ defined by $f(u_1, \ldots, u_N) := r(u_1, \ldots, u_{N-1})$ is continuous and odd.

5. Contraction property

The degree is independent of the coordinate system.

Proposition D.19. *Let* $g : \mathbb{R}^N \to \mathbb{R}^N$ *be a diffeomorphism. Then*

$$\deg(f, U) = \deg(g \circ f \circ g^{-1}, g(u)).$$

Proof. By using the Weierstrass approximation theorem and the Sard theorem, we can assume that $f \in C^2(U, \mathbb{R}^N)$ and that 0 is a regular value of f. It suffices then to use the definition of the degree. $\square$

Theorem D.20. (Contraction property). *Let* $f \in C(\bar{U}, \mathbb{R}^N)$ *be such that* $0 \notin f(\partial U)$. *If there exists a subspace* Y *of* $\mathbb{R}^N$ *such that* $(\mathrm{id} - f)(U) \subset Y$ *then*

$$\deg(f, U) = \deg(f\big|_{U \cap Y}, U \cap Y).$$

Proof. By the preceding proposition, we can assume that $Y = \mathbb{R}^M$. By using the Weierstrass approximation theorem and the Sard theorem, we can also assume that $f \in C^2(U, \mathbb{R}^N)$ and that 0 is a regular value of f. If $f(y, 0) = 0$, then

$$\det(f'(y, 0)) = \det \begin{pmatrix} \partial_y f(y, 0) & \partial_z f(y, 0) \\ 0 & \mathrm{id}\big|_{\mathbb{R}^{N-M}} \end{pmatrix} = \det(f\big|_{U \cap \mathbb{R}^M})'(y)$$

and it suffices to use the definition of the degree. $\square$

Bibliography

[1] Alama S. and Li Y.Y., Existence of solutions for semilinear elliptic equations with indefinite linear part, *J. Diff. Eq.* **96** (1992) 89-115.

[2] Ambrosetti A., Critical points and nonlinear variational problems, *Bull. Soc. Math. France* **120** (1992) Mémoire n⁰ 49.

[3] Ambrosetti A., Brézis H. and Cerami G., Combined effects of concave and convex nonlinearities in some elliptic problems, *J. Funct. Anal.* **122** (1994) 519-543.

[4] Ambrosetti A. and Rabinowitz P.H., Dual variational methods in critical point theory and applications, *J. Funct. Anal.* **14** (1973) 349-381.

[5] Aubin Th., Problèmes isopérimétriques et espaces de Sobolev, *J. Diff. Geom.* **11** (1976) 573-598.

[6] Aubin J.P. and Ekeland I., *Applied nonlinear analysis*, Wiley, New York, 1984.

[7] Bartsch T., Infinitely many solutions of a symmetric Dirichlet problem, *Nonlinear Analysis*, TMA **20** (1993) 1205-1216.

[8] Bartsch T., Topological Methods for variational problems with symmetrics, *Lecture Notes in Math.* **1560**, Springer, Berlin, 1993.

[9] Bartsch T. and Willem M., Infinitely many nonradial solutions of a Euclidean scalar field equation, *J. Funct. Anal.* **117** (1993) 447-460.

[10] Bartsch T. and Willem M., Infinitely many radial solutions of a semilinear elliptic problem on $\mathbb{R}^N$, *Arch. Rat. Mech. Anal.* **124** (1993) 261-276.

[11] Bartsch T. and Willem M., Periodic solutions of non-autonomous Hamiltonian systems with symmetries, *J. Reine Angew. Math.* **451** (1994) 149-159.

[12] Bartsch T. and Willem M., On an elliptic equation with concave and convex nonlinearities, *Proc. Amer. Math. Soc.* **123** (1995) 3555-3561.

[13] Ben-Naoum A.K., Troestler C. and Willem M., Extrema problems with critical Sobolev exponents on unbounded domains, *Nonlinear Analysis*, TMA **26** (1996) 823-833.

[14] Benci V. and Cerami G., Positive solutions of semilinear elliptic problems in exterior domains, *Arch. Rat. Mech. Anal.* **99** (1987) 283-300.

[15] Benci V. and Rabinowitz P.H., Critical point theorems for indefinite functionals, *Inv. Math.* **52** (1979) 241-273.

[16] Berestycki H. and Lions P.L., Nonlinear scalar field equations, *Arch. Rat. Mech. Anal.* **82** (1983) 313-376.

[17] Besov, O.V., Ilin V.P. and Nikolki S.M., *Integral Representation of functions and imbeddings theorems*, Vol. I, Wiley, New York, 1978.

[18] Bianchi G., Chabrowski J. and Szulkin A., On symmetric solutions of an elliptic equation with a nonlinearity involving critical Sobolev exponent, *Nonlinear Analysis*, TMA **25** (1995) 41-59.

[19] de Bouard A. and Saut J.C., Sur les ondes solitaires des équations de Kadomtsev-Petviashvili, *C. R. Acad. Sciences Paris* **320** (1995) I315-I328, and Solitary waves of generalized Kadomtsev-Petviashvili equations, Prépublications de l'Université de Paris-Sud, 34-66, 1994.

[20] Brézis H., *Analyse fonctionnelle*, Masson, Paris, 1983.

[21] Brézis H., Elliptic equations with limiting Sobolev exponents - The impact of topology, *Comm. Pure Appl. Math.* **39** (1986) 517-539.

[22] Brézis H. and Coron J.M., Convergence of solutions of *H*-systems or how to blow bubbles, *Arch. Rat. Mech. Anal.* **89** (1985) 21-56.

[23] Brézis H., Coron J.M. and Nirenberg L., Free vibrations for a nonlinear wave equation and a theorem of P. Rabinowitz, *Comm. Pure Appl. Math.* **33** (1980) 667-689.

[24] Brézis H. and Kato T., Remarks on the Schrödinger operator with singular complex potentials, *J. Math. Pures et Appl.* **58** (1979) 137-151.

[25] Brézis H. and Lieb E., A relation between pointwise convergence of functions and convergence of functionals, *Proc. Amer. Math. Soc.* **88** (1983) 486-490.

[26] Brézis H. and Nirenberg L., Positive solutions of nonlinear elliptic equations involving critical Sobolev exponents, *Comm. Pure Appl. Math.* **36** (1983) 437-477.

[27] Brézis H. and Nirenberg L., Remarks on finding critical points, *Comm. Pure Appl. Math.* **64** (1991) 939-963.

[28] Buffoni B., Jeanjean L. and Stuart C.A., Existence of nontrivial solutions to a strongly indefinite semilinear equation, *Proc. Amer. Math. Soc.* **119** (1993) 179-186.

[29] Capozzi A., Fortunato D. and Palmieri G., An existence result for nonlinear elliptic problems involving critical Sobolev exponent, Ann. Inst. Henri Poincaré, *Analyse Non linéaire* **2** (1985) 463-470.

[30] Chang K.C., *Infinite dimensional Morse theory and applications to differential equations*, Birkhaüser, Boston, 1992.

[31] Coti-Zelati V. and Rabinowitz P., Homoclinic type solutions for a semilinear elliptic PDE on $\mathbb{R}^N$, *Comm. Pure Appl. Math.* **45** (1992) 1217-1269.

[32] Ding W.Y. and Ni W.M., On the existence of positive entire solutions of a semilinear elliptic equation, *Arch. Rat. Math. Anal.* **31** (1986) 283-308.

[33] Dugundji J., An extension of Tietze's theorem, *Pac. J. Math.* **1** (1951) 353-367.

[34] Ekeland I., On the variational principle, *J. Math. Anal. Appl.* **47** (1974) 324-353.

[35] Ekeland I., *Convexity methods in Hamiltonian mechanics*, Springer, Berlin, 1990.

[36] Fournier G., Lupo D., Ramos M. and Willem M., Limit relative category and critical point theory, *Dynamics Reported* **3** (1994) 1-24.

[37] Garcia Azorero J. and Peral Alonso I., Multiplicity of solutions for elliptic problems with critical exponent or with a nonsymmetric term, *Trans. Amer. Math. Soc.* **323** (1991) 877-895.

[38] Ghoussoub N., *Duality and perturbation methods in critical point theory*, Cambridge University Press, Cambridge, 1993.

[39] Gromes W., Ein einfacher Beweis des Satzes von Borsuk, *Math. Zeitschrift* **178** (1981) 399-400.

[40] Heinz H., Küpper T. and Stuart C., Existence and bifurcation of solutions for nonlinear perturbations of the periodic Schrödinger equation, *J. Diff. Eq.* **100** (1992) 341-354.

[41] Hofer H. and Wysocki, First order elliptic systems and the existence of homoclinic orbits in Hamiltonian systems, *Math. Annalen* **228** (1990) 483-503.

[42] Hofer H. and Zehnder E., *Symplectic invariants and Hamiltonian dynamics*, Birkhauser, Basel, 1994.

[43] Jeanjean L., Solutions in spectral gaps for a nonlinear equation of Schrödinger type, *J. Diff. Eq.* **112** (1994) 53-80.

[44] Kavian O., *Introduction à la théorie des points critiques et applications aux problèmes elliptiques*, Springer, Heidelberg, 1993.

[45] Kryszewski W. and Szulkin A., On a semilinear Schrödinger equation with indefinite linear part, preprint, 1996.

[46] Lazzo M., Solutions positives multiples pour une équation elliptique non linéaire avec l'exposant critique de Sobolev, *C. R. Acad. Sci. Paris* **314** (1992) I61-I64.

[47] Li S., Some existence theorems of critical points and applications, IC/86/90 Report, ICTP, Trieste.

[48] Li S. and Willem M., Applications of local linking to critical point theory, *J. Math. Anal. Appl.* **189** (1995) 6-32.

[49] Lions P.L., Symétrie et compacité dans les espaces de Sobolev, *J. Funct. Anal.* **49** (1982) 315-334.

[50] Lions P.L., The concentration-compactness principle in the calculus of variations. The locally compact case. Ann. Inst. Henri Poincaré, *Analyse Non Linéaire* **1** (1984) 109-145 and 223-283.

[51] Lions P.L., The concentration-compactness principle in the calculus of variations. The limit case. *Rev. Mat. Ibero americana* **1** (1985) 145-201 and **2** (1985) 45-121.

[52] Lions P.L., Symmetrics and the concentration compactness method, in *Nonlinear Variational Problems*, Pitman, London, 1985, 47-56.

[53] Lopes O., Radial symmetry of minimizers for some translation and rotation invariant functionals, *J. Diff. Eq.* **124** (1966) 378-388.

[54] Lopes O., Radial and nonradial minimizers for some radially symmetric functionals, preprint.

[55] Lusternik L. and Schnirelman L., *Méthodes topologiques dans les problèmes variationnels*, Hermann, Paris, 1934.

[56] Mawhin J., *Problèmes de Dirichlet variationnels non linéaires*, Presses de l'Université de Montréal, Montréal, 1987.

[57] Mawhin J. and Willem M., *Critical point theory and Hamiltonian systems*, Springer, New York, 1989.

[58] Nagumo M., A note on the theory of degree of mapping in Euclidean spaces, *Osaka Math. J.* **4** (1952) 1-5.

[59] Nehari Z., On a class of nonlinear second-order differential equations, *Trans. Amer. Math. Soc.* **95** (1960) 101-123.

[60] Nehari Z., Characteristic values associated with a class of nonlinear second-order differential equations, *Acta Math.* **105** (1961) 141-175.

[61] Nehari Z., On a nonlinear differential equation arising in nuclear physics, *Proc. Roy. Irish Acad. Sect.* **A 62** (1963) 117-135.

[62] Palais R., Lusternik-Schnirelman theory on Banach manifolds, *Topology* **5** (1966) 115-132.

[63] Palais R.S., The principle of symmetric criticality, *Comm. Math. Phys.* **69** (1979) 19-30.

[64] Pohozaev S., Eigenfunctions of the equation $\Delta u + \lambda f(u) = 0$, *Soviet. Math. Dokl.* **6** (1965) 1408-1411.

[65] Pucci P. and Serrin J., A general variational identity, *Indiana Univ. Math. J.* **35** (1986) 681-703.

[66] Ramos M., *Teoremas de enlace na teoria dos pontos criticas*, Universidade de Lisboa, Departamento de Matematica, Lisboa, 1993.

[67] Rabinowitz P.H., Periodic solutions of Hamiltonian systems, *Comm. Pure Appl. Math.* **31** (1978) 157-184.

[68] Rabinowitz P.H., Some critical point theorems and applications to semilinear elliptic partial differential equations, Ann. Scuola Normale Sup. Pisa, *Classe Scienza* **4** (1978) 215-223.

[69] Rabinowitz P.H., Some minimax theorems and applications to nonlinear partial differential equations, in *Nonlinear analysis: A collection of paper in honor of Erich Röthe*, Academic Press, New York, 1978, 161-177.

[70] Rabinowitz P.H., *Minimax methods in critical point theory with applications to differential equations*, Amer. Math. Soc., Providence, 1986.

[71] Rabinowitz P.H., A note on a semilinear elliptic equation on $\mathbb{R}^N$, in *Nonlinear Analysis: A Tribute in Honour of G. Prodi, Quaderni Sc. Norm. Sup. Pisa*, Pisa, 1991, 307-317.

[72] Rabinowitz P.H., On a class of nonlinear Schrödinger equations, *ZAMP* **43** (1992) 270-291.

[73] Reeken M., Stability of critical points under small perturbations, *Manuscripta Math.* **7** (1972) 387-411.

[74] Rellich F., Darstellung der eigenwerte von $\Delta u + \lambda u = 0$ durch ein randintegral, *Math. Zeit.* **46** (1940) 635-636.

[75] Rey O., A multiplicity result for a variational problem with lack of compactness, *Nonlinear Analysis*, TMA **13** (1989) 1241-1249.

[76] Sanchez L., *Metodos da teoria de pontos criticas*, Universidade de Lisboa, Departamento de Matematica, Lisboa, 1993.

[77] Sard A., The measure of the critical values of differentiable maps, *Bull. Amer. Math. Soc.* **48** (1942) 883-890.

[78] Schwartz L., *Cours d'analyse*, Hermann, Paris, 1991-1994.

[79] Stuart C., *Bifurcation into spectral gaps*, Société Mathématique de Belgique, Bruxelles, 1995.

[80] Strauss W.A., Existence of solitary waves in higher dimensions, *Comm. Math. Phys.* **55** (1977) 149-162.

[81] Struwe M., *Variational methods*, Springer, Berlin, 1990.

[82] Struwe M., A global compactness result for elliptic boundary value problems involving limiting nonlinearities, *Math. Z.* **187** (1984) 511-517.

[83] Szulkin A., Ljusternik-Schnirelmann theory on $\mathcal{C}^1$-manifolds, Ann. Inst. Henri Poincaré, *Analyse Non Linéaire* **5** (1988) 119-139.

[84] Szulkin A., A relative category and applications to critical point theory for strongly indefinite functionals, *Nonlinear Analysis, TMA* **15** (1990) 725-739.

[85] Talenti G., Best constants in Sobolev inequality, *Annali di Mat.* **110** (1976) 353-372.

[86] Troestler C. and Willem M., *Nontrivial solution of a semilinear Schrödinger equation*, Communications in Partial Differential Equations, to appear.

[87] Van der Vorst R.C.A.M., Variational identities and applications to differential systems, *Arch. Rat. Mech. Anal.* **116** (1991) 375-398.

[88] Wang Z.Q., On a superlinear elliptic equation, Ann. Inst. Henri Poincaré, *Analyse Non Linéaire* **8** (1991) 43-57.

[89] Willem M., Lectures on critical point theory, *Trabalho de Math.* **199**, Fundação Univ. Brasilia, Brasilia, 1983.

[90] Willem M., *Analyse harmonique réelle*, Hermann, Paris, 1995.

Index of Notations

We use the following notations:

Ω : domain of $\mathbb{R}^N$,

$|u|_p := \left(\int_\Omega |u(x)|^p dx \right)^{1/p}$,

$\mathcal{D}(\Omega) := \{ u \in \mathcal{C}^\infty(\Omega) : \text{supp } u \text{ is a compact subset of } \Omega \}$.

$H^1(\mathbb{R}^N), \mathcal{D}^{1,2}(\mathbb{R}^N), H_0^1(\Omega), \mathcal{D}_0^{1,2}(\Omega)$: Sobolev spaces 1.7.

$2^* := \infty, \quad N = 1, 2,$

$\quad := 2N/(N-2), \quad N \geq 3.$

$\overset{o}{A}$: interior of A,

$\bar{A}$: closure of A,

$B(x, r)$: open ball with center x and radius r,

$B[x, r]$: closed ball with center x and radius r.

We denote by $\to$ (resp. $\rightharpoonup$) the strong (resp. weak) convergence.

Let φ be a real function defined on a normed space X and let S be a subset of X,

$$\varphi^d := \{ u \in X : \varphi(u) \leq d \},$$
$$S_\delta := \{ u \in X : \text{dist}(u, S) \leq \delta \}.$$

Index

Progress in Nonlinear Differential Equations and Their Applications

Editor

Haim Brezis
Département de Mathématiques
Université P. et M. Curie
4, Place Jussieu
75252 Paris Cedex 05
France
and
Department of Mathematics
Rutgers University
New Brunswick, NJ 08903
U.S.A.

Progress in Nonlinear Differential Equations and Their Applications is a book series that lies at the interface of pure and applied mathematics. Many differential equations are motivated by problems arising in such diversified fields as Mechanics, Physics, Differential Geometry, Engineering, Control Theory, Biology, and Economics. This series is open to both the theoretical and applied aspects, hopefully stimulating a fruitful interaction between the two sides. It will publish monographs, polished notes arising from lectures and seminars, graduate level texts, and proceedings of focused and refereed conferences.

We encourage preparation of manuscripts in some form of TeX for delivery in camera-ready copy, which leads to rapid publication, or in electronic form for interfacing with laser printers or typesetters.

Proposals should be sent directly to the editor or to: Birkhäuser Boston, 675 Massachusetts Avenue, Cambridge, MA 02139